NELSON M. BLACHMAN is a senior scientist with Sylvania Electronic Systems, Mountain View, California, serving as a consultant on communication theory. He holds a B.S. degree in physics from Case Institute of Technology, and an M.A. degree in physics and a Ph.D. degree in engineering sciences and applied physics from Harvard University. He was associated at Harvard with the Underwater Sound Laboratory and the Cruft Laboratory.

Dr. Blachman has been a member of the staff of the Brookhaven National Laboratory, where he was involved in the development of the Cosmotron, and of the staff of the Office of Naval Research in both Washington and London. A frequent contributor to scientific journals, he has taught in the off-campus programs of the Universities of Maryland and California and, under a Fulbright grant, spent a year teaching communication theory in Madrid.

Dr. Blachman is a fellow of the Institute of Electrical and Electronics Engineers and a member of the International Scientific Radio Union, the Institute of Mathematical Statistics, the Mathematical Association of America, and Sigma Xi.

*noise and its effect on communication*

McGRAW-HILL ELECTRONIC SCIENCES SERIES

EDITORIAL BOARD
Ronald Bracewell Colin Cherry Willis W. Harman
Edward W. Herold John G. Linvill Simon Ramo John G. Truxal

ABRAMSON Information theory and coding
BLACHMAN Noise and its effect on communication
BREMER Superconductive devices
BROXMEYER Inertial navigation systems
GILL Introduction to the theory of finite-state machines
HANCOCK AND WINTZ Signal detection theory
HUELSMAN Circuits, matrices, and linear vector spaces
KELSO Radio ray propagation in the ionisphere
MERRIAM Optimization theory and the design of feedback control systems
MILSUM Biological control systems analysis
NEWCOMB Linear multiport synthesis
PAPOULIS The fourier integral and its applications
STEINBERG AND LEQUEUX (TRANSLATOR R. N. BRACEWELL) Radio astronomy

# NOISE and its effect on COMMUNICATION

*nelson m. blachman* Sylvania Electronic Systems
Mountain View, California

LIBRARY
BRYAN COLLEGE
DAYTON, TN. 37321

*mcgraw-hill book company* New York San Francisco St. Louis
London Toronto Sydney

58863

NOISE AND ITS EFFECT
ON COMMUNICATION

Copyright © 1966 by McGraw-Hill, Inc. All rights reserved. This book or parts thereof, may not be reproduced in any form without permission of the publishers.

Printed in the United States of America

Library of Congress catalog card number: 66-19283

234567890 MP 72106987

05497

# PREFACE

The noise with which this book deals is mainly the inevitable gaussian noise that arises in every electrical circuit and limits the useful sensitivity of electronic equipment, in which it may produce a hissing or roaring sound. The approach used and some of the results obtained are general enough to apply to other types of noise as well, such as lightning discharges and interchannel interference, but the gaussian case is the most important illustration and is the easiest to describe and to analyze. Accordingly, Chapter 1 discusses the univariate, bivariate, and multivariate normal (gaussian) distributions, their moments, and related probability distributions.

Chapter 2 takes up random processes, both gaussian and nongaussian, and focuses attention on the most interesting class of such processes or noises, namely, those that are ergodic. In Chapter 3 the Wiener-Khinchin theorem, relating the power spectrum of an ergodic random process to its correlation function, is derived and is applied to a variety of random noises and random signals, including those used for AM and for FM communication, and the physical reasons are presented for the nonexistence of the limit by which the power spectral density has often been defined. The first part of the book, on noise and random signals, ends with Chapter 4, dealing with their statistics in the narrow-band case, which is the most important for radio communication. These statistics are exploited to obtain many useful results very simply.

Part II deals with the effect of noise upon signals in various nonlinear devices, such as AM and FM demodulators, limiters, harmonic generators, and unintended nonlinearities. The output signal and noise are studied in Chapters 5 and 6, and in Chapter 7 the effect of the noise is determined on the probability of error in the detection of coherent, incoherent, and noise-like signals, as in the case of telegraphy and teletype, or radio astronomy.

The third and last part of the book presents the fundamentals of information theory, generally following the lines of Shannon's original papers. Chapter 8 deals with the measure of information and its applications in the discrete case, including cryptography and the redundancy of English, as well as the basic coding theorems, and Chapter 9 treats communication through continuous, noisy channels.

The aim of the book is to present the basic tools of statistical communication theory in a strictly logical order and to show some of their applications. These applications of probability concepts and statistical methods principally concern communication processes, like radio and telephony, but the approach presented here is applicable as well to radar, to control systems, and to other situations in which noise limits system performance. The book does not delve into the physical origins of the noise, which are best treated[1] as an aspect of the electronic devices and phenomena that produce them.

The mathematical exposition is mainly heuristic, developing and exploiting physical insight and geometric methods as fully as possible in order to avoid needlessly complicated calculations, whose results might be difficult to interpret and to apply. This is particularly true of the second half of the book, and there will therefore be readers who find the last three chapters insufficiently rigorous, that is, insufficiently convincing. However, most readers are likely to find the heuristic approaches more persuasive and easier to apply than the more involved treatments appearing in more advanced works, some of which are cited in the footnotes. The solution of communication-theoretical problems requires a knowledge not only of the basic facts, principles, and techniques of the field but also of the art of asking the right questions, making suitable approximations, and avoiding blind alleys and unnecessary complication. The latter can be learned only through experience, and this book is intended to provide some of that experience.

The book is written for engineers, physicists, and mathematicians, whom it should enable to solve a wide variety of practical problems with convincing simplicity. It should also prepare them for more advanced study so that they can take up the current journal literature, Middleton's encyclopedic work,[2] or more specialized books.

The reader is assumed to have an elementary knowledge of probability theory and Fourier integral analysis, such as would be provided

[1] W. R. Bennett, "Electrical Noise," McGraw-Hill Book Company, New York, 1960; J. L. Lawson and G. E. Uhlenbeck, "Threshold Signals," M.I.T. Rad. Lab. Series No. 24, chaps. 4 and 5, McGraw-Hill Book Company, New York, 1950; J. C. Hancock, "An Introduction to the Principles of Communication Theory," McGraw-Hill Book Company, New York, 1961; D. A. Bell, "Electrical Noise," D. Van Nostrand Company, Inc., Princeton, N. J., 1960.

[2] D. Middleton, "An Introduction to Statistical Communication Theory," McGraw-Hill Book Company, New York, 1960.

by Woodward's, Bracewell's, or Hancock's books.[1] He will also need to know a little about the common types of communication systems, linear filters, and matrix notation. Those having professional experience with communication systems will find that they are already well motivated and that their physical feeling for the problems taken up here is put to good use.

Some 189 problems are included, in part for practice in using the principles and techniques presented, in part to furnish examples and short proofs that are needed in the text, and in part to give additional results that would otherwise not easily fit into the book.

This book had its origin in a course in statistical communication theory which I taught in the Engineering Extension program of the University of California in 1961–1962 and which I offered again in 1963–1964. It was more or less completed during the academic year 1964–1965 while I was teaching the same course, this time in Spanish, at the University of Madrid and the Escuela Técnica Superior de Ingenieros de Telecomunicación, Madrid, under the auspices of the Fulbright program for educational and cultural exchange.

The book contains sufficient material for a full year's course. Where specialized courses on detection theory and information theory are available, however, the first six chapters can serve for a one-semester course on noise and random signals and their processing.

I am grateful to Robert Price, Robert G. Gallager, and David Middleton for their very helpful comments on the manuscript of this book and to my wife for her patience throughout its writing.

NELSON M. BLACHMAN

[1] P. M. Woodward, "Probability and Information Theory with Applications to Radar," chaps. 1 and 2, Pergamon Press, Ltd., London, and McGraw-Hill Book Company, New York, 1953; Ron Bracewell, "The Fourier Transform and Its Applications," McGraw-Hill Book Company, New York, 1965; J. C. Hancock, *op. cit.* The necessary background and a good deal more are provided by W. W. Harman, "Principles of the Statistical Theory of Communication," McGraw-Hill Book Company, New York, 1963.

# CONTENTS

# PART I

# Statistical Properties of Noise and Random Signals

# chapter 1 THE NORMAL DISTRIBUTION

The interference arising in communication systems takes many forms, but we shall concentrate on one very common and relatively tractable kind—gaussian noise—whose univariate, bivariate, and multivariate probability distributions (describing the values taken by the process at one, two, and many instants of time, respectively) all have a particular form known as *normal* or *gaussian*. The special importance of the normal distribution is explained by the central limit theorem of probability theory, which asserts that the distribution of the sum of a large number of independent random variables whose variances are small compared to their sum tends toward the normal distribution with mean and variance given by the sum of the means and the sum of the variances of the summands, respectively. Thus, since electrical noise often results from the superposition of the effects of many electrons, it often takes the form of gaussian noise.

## THE UNIVARIATE NORMAL DISTRIBUTION

A univariate normal distribution of, say, $x$ is characterized completely by two parameters, the first moment or mean

$$\mu = \mathrm{E}\,\{x\} = \int_{-\infty}^{\infty} xp(x)\,dx \tag{1-1}$$

and the variance or second moment about the mean

$$\sigma^2 = \mathrm{E}\,\{(x - \mu)^2\} = \int_{-\infty}^{\infty} (x - \mu)^2 p(x)\,dx. \tag{1-2}$$

Here E is an operator which takes the expectation or average over the probability distribution with density function $p(x)$; that is, $p(x)\,dx$ is the probability that $x$ falls within an interval of width $dx$. The square root $\sigma$

of the variance of $x$ is called its standard deviation. The univariate normal probability density function $p(x)$ or, more fully, $p(x|\mu,\sigma)$ has the form (Fig. 1-1)

$$p(x|\mu,\sigma) = \frac{1}{\sqrt{2\pi}\,\sigma} \exp\left[-\frac{(x-\mu)^2}{2\sigma^2}\right]. \tag{1-3}$$

**Problem 1-1.** Verify that (1-3) is a probability density, i.e., that it is nonnegative and its integral over all $x$ is unity. [HINT: Write the square of this integral as a double integral over $x$ and $y$. Then change to polar coordinates with origin at the point $(\mu,\mu)$.]

**Problem 1-2.** Verify that the distribution with density function (1-3) has mean (1-1) and variance (1-2).

**Problem 1-3.** Show that the second moment E $\{x^2\}$ of the distribution (1-3) is $\sigma^2 + \mu^2$ by showing that the second moment of *any* distribution is the sum of its variance plus the square of its mean.

**Moments.** It is often useful to know the higher "central moments" E $\{(x-\mu)^k\}$ of the normal distribution. Since (1-3) is symmetric about its mean, they vanish for all odd $k$. Through repeated integration by parts we find that, for even $k$,

$$\begin{aligned} \mathrm{E}\,\{(x-\mu)^k\} &= \int_{-\infty}^{\infty} \frac{(x-\mu)^k}{\sqrt{2\pi}\,\sigma} \exp\left[-\frac{(x-\mu)^2}{2\sigma^2}\right] dx \\ &= 1 \cdot 3 \cdot 5 \cdot \cdot \cdot (k-1)\sigma^k. \end{aligned} \tag{1-4}$$

In particular, the second central moment is the variance $\sigma^2$, and the fourth central moment E $\{(x-\mu)^4\}$ is three times the square of the variance.

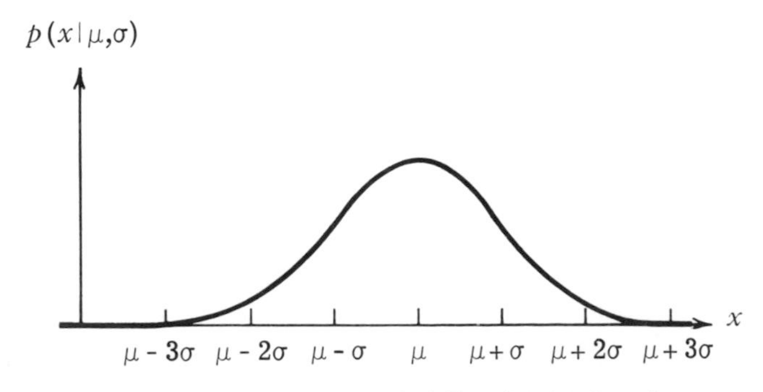


*Figure 1-1.* The univariate normal probability density function.

**Cumulative Distribution Function.** The *standard normal distribution* is the case of zero mean and unit variance, with density function $p(x|0,1)$. If $x$ is normal with mean $\mu$ and standard deviation $\sigma$, then $(x - \mu)/\sigma$ is a standard normal random variable. For finding error probabilities we shall have occasion to make use of the cumulative distribution function of the standard normal distribution, i.e., the probability that a standard normal random variable does not exceed $x$, which is the error function

$$\text{erf } x = \frac{1}{\sqrt{2\pi}} \int_{-\infty}^{x} e^{-\frac{1}{2}t^2}\, dt \tag{1-5}$$

(Fig. 1-2), sometimes denoted by $\Phi(x)$. The probability that a normal random variable with mean $\mu$ and standard deviation $\sigma$ does not exceed $x$ is thus erf $[(x - \mu)/\sigma]$. Variants of this error function are found in many

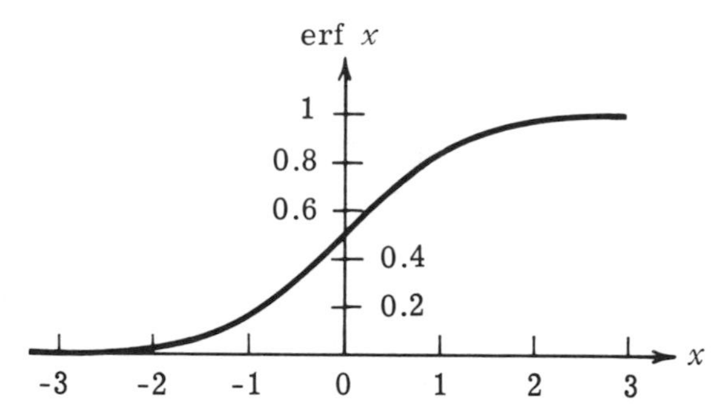


*Figure 1-2.* The error function.

books of mathematical tables, but the tables are generally inadequate for extreme values of $x$.

**Asymptotic Series for the Error Function.** In order to approximate erf $x$ for very negative $x$, we substitute $x - u$ for $t$ in (1-5), getting

$$\begin{aligned} \text{erf } x &= \frac{1}{\sqrt{2\pi}} \int_0^{\infty} e^{-\frac{1}{2}(x-u)^2}\, du \\ &= \frac{e^{-\frac{1}{2}x^2}}{\sqrt{2\pi}} \int_0^{\infty} e^{-\frac{1}{2}u^2} e^{xu}\, du. \end{aligned}$$

Now by Taylor's theorem, for any real $u$, $e^{-\frac{1}{2}u^2}$ lies between the successive partial sums of the power series

$$1 - \frac{\frac{1}{2}u^2}{1!} + \frac{(\frac{1}{2}u^2)^2}{2!} - \frac{(\frac{1}{2}u^2)^3}{3!} + \cdots.$$

Hence, for $x < 0$, we see, on using $n + 1$ terms of this series, that erf $x$ lies between the values taken by

$$\begin{aligned}
&\frac{e^{-\frac{1}{2}x^2}}{\sqrt{2\pi}} \int_0^\infty \left[1 - \frac{u^2}{2} + \frac{u^4}{2 \cdot 4} - \cdots \pm \frac{u^{2n}}{2 \cdot 4 \cdot 6 \cdots (2n)}\right] e^{xu}\, du \\
&= \frac{e^{-\frac{1}{2}x^2}}{-\sqrt{2\pi}\, x} \int_0^\infty \left[1 - \frac{v^2}{2x^2} + \frac{v^4}{2 \cdot 4x^4} - \cdots \pm \frac{v^{2n}/x^{2n}}{2 \cdot 4 \cdot 6 \cdots (2n)}\right] e^{-v}\, dv \\
&= \frac{e^{-\frac{1}{2}x^2}}{-\sqrt{2\pi}\, x} \left[1 - \frac{1}{x^2} + \frac{1 \cdot 3}{x^4} - \cdots \pm \frac{1 \cdot 3 \cdot 5 \cdots (2n-1)}{x^{2n}}\right] \qquad (1\text{-}6)
\end{aligned}$$

for even and for odd $n$.

Here we have made use of the integral

$$\int_0^\infty v^s e^{-v}\, dv = \Gamma(s + 1) \qquad (1\text{-}7)$$

defining the gamma function for $s > -1$. When $s$ is a positive integer, (1-7) is $s! = 1 \cdot 2 \cdot 3 \cdots s$, and it will be convenient throughout this book to use $s!$ to mean $\Gamma(s + 1)$ even when $s$ is not an integer. Thus, setting $s = -\frac{1}{2}$ and $v = \frac{1}{2}x^2$ in (1-7), we get the integral arising in Prob. 1-1, from which we have $\Gamma(\frac{1}{2}) = (-\frac{1}{2})! = \sqrt{\pi}$. Integrating (1-7) by parts, we find $(s + 1)! = (s + 1)s!$, whence $\frac{1}{2}! = \frac{1}{2}\sqrt{\pi}$, $\frac{3}{2}! = \frac{3}{2} \cdot \frac{1}{2}\sqrt{\pi}$, and so on.

For very negative $x$, the successive terms of the series (1-6) decrease very rapidly at first, thus permitting accurate evaluation of erf $x$. However, if $n$ is made infinite, the series will diverge for all $x$. An expansion like (1-6) is called *asymptotic.*[1]

Because the sum of the areas in Fig. 1-1 to the left and to the right of any $x$ is unity and because the normal density function (1-3) is symmetric about its mean, we have erf $x$ + erf $-x$ = erf $\infty$ = 1. Thus, for $x > 0$, we have the asymptotic expansion

$$\operatorname{erf} x \sim 1 - \frac{e^{-\frac{1}{2}x^2}}{\sqrt{2\pi}\, x} \left[1 - \frac{1}{x^2} + \frac{1 \cdot 3}{x^4} - \cdots \pm \frac{1 \cdot 3 \cdot 5 \cdots (2n-1)}{x^{2n}}\right], \qquad (1\text{-}8)$$

which, for $n = 0, 1, 2, \ldots$, leaves an error smaller than the first omitted term. Thus (1-8) is useful for large positive $x$.

**Problem 1-4.** Use (1-6) and (1-8) to find the probability that a normal random variable will depart from its mean by more than five standard deviations, and check the result by reference to a table of the error function.

[1] A. Erdélyi, "Asymptotic Expansions," Dover Publications, Inc., New York, 1956.

**Problem 1-5.** By making use of (1-7) show that the $k$th absolute moment of the standard normal distribution is $\mathrm{E}\{|x|^k\} = 2^{\frac{1}{2}k}\left(\frac{k-1}{2}\right)!/\sqrt{\pi}$, and show that this result is in agreement with (1-4).

## THE BIVARIATE NORMAL DISTRIBUTION

The bivariate normal distribution[1] of $x_1$ and $x_2$ with means

$$\mu_1 = \mathrm{E}\{x_1\} \qquad \text{and} \qquad \mu_2 = \mathrm{E}\{x_2\}, \tag{1-9}$$

respectively, and variances

$$\sigma_1^2 = \mathrm{E}\{(x_1 - \mu_1)^2\} \qquad \text{and} \qquad \sigma_2^2 = \mathrm{E}\{(x_2 - \mu_2)^2\} \tag{1-10}$$

is characterized by one additional parameter relating these two random variables, namely, their coefficient of correlation

$$\varrho = \frac{\mathrm{E}\{(x_1 - \mu_1)(x_2 - \mu_2)\}}{\sigma_1\sigma_2}. \tag{1-11}$$

Since the square of a real quantity cannot be negative, the average value of the square cannot be negative, and we have the two inequalities

$$\mathrm{E}\left\{\left(\frac{x_1 - \mu_1}{\sigma_1} \pm \frac{x_2 - \mu_2}{\sigma_2}\right)^2\right\} \geq 0,$$

from which it follows that $-1 \leq \varrho \leq 1$.

In terms of these five parameters the bivariate normal probability density function is

$$p(x_1,x_2|\mu_1,\mu_2,\varrho,\sigma_1,\sigma_2) = \frac{1}{2\pi\sigma_1\sigma_2\sqrt{1-\varrho^2}} \exp\left[-\frac{\dfrac{(x_1-\mu_1)^2}{2\sigma_1^2} - \varrho\dfrac{x_1-\mu_1}{\sigma_1}\dfrac{x_2-\mu_2}{\sigma_2} + \dfrac{(x_2-\mu_2)^2}{2\sigma_2^2}}{1-\varrho^2}\right], \tag{1-12}$$

that is, the probability that the point $(x_1,x_2)$ falls in a small rectangle of dimensions $dx_1$, $dx_2$ is $dx_1\,dx_2$ times (1-12). Because $\varrho^2 \leq 1$, the exponent in (1-12) can never be positive, and it attains its maximum value, zero, at the point $(\mu_1,\mu_2)$. The contours of constant probability density (1-12) in the $x_1x_2$ plane are the loci where this exponent is constant; they are similar, coaxial ellipses (Fig. 1-3), the inclination of their major axis being $\alpha = \frac{1}{2}\arctan 2\varrho\sigma_1\sigma_2/(\sigma_1^2 - \sigma_2^2)$.

[1] H. Cramér, "Methods of Mathematical Statistics," p. 287, Princeton University Press, Princeton, N.J., 1946; W. Feller, "An Introduction to Probability Theory and Its Applications," vol. 2, p. 69, John Wiley & Sons, Inc., New York, 1966.

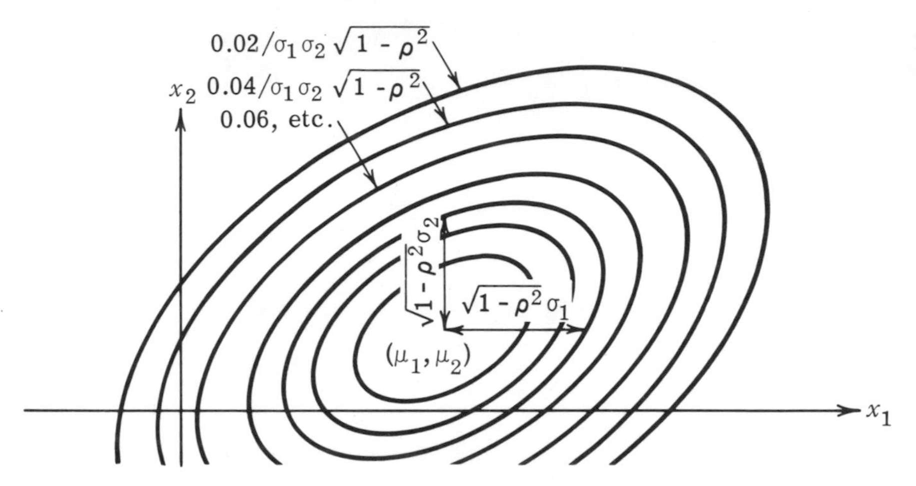


*Figure 1-3.* Contours of the bivariate normal probability density function with $\varrho > 0$. With $\varrho < 0$, the major axis of the ellipses has a negative slope.

**Problem 1-6.** Verify (1-9), (1-10), and (1-11) for the density function (1-12), and show that its integral with respect to $x_2$ from $-\infty$ to $\infty$ is the univariate normal density function $p(x_1|\mu_1,\sigma_1)$ given by (1-3).

**Problem 1-7.** Show that, if $x_1$ and $x_2$ are jointly normal with mean zero and correlation coefficient $\varrho$, the probability that both have the same sign is $\frac{1}{2} + \frac{1}{\pi} \arcsin \varrho$. (HINT: Let $x_1 = r \cos \theta$, $x_2 = r \sin (\theta + \phi)$, with $\sin \phi = \varrho$ and $dx_1\, dx_2 = r \cos \phi \, d\theta \, dr$.)

**Conditional Distribution.** Since the joint density function $p(x_1,x_2)$ of $x_1$ and $x_2$ is the product of the density function $p(x_1)$ of $x_1$ times the conditional density function $p(x_2|x_1)$ of $x_2$ for a given value of $x_1$, the latter is

$$\begin{aligned} p(x_2|x_1,\mu_1,\mu_2,\varrho,\sigma_1,\sigma_2) &= \frac{p(x_1,x_2|\mu_1,\mu_2,\varrho,\sigma_1,\sigma_2)}{p(x_1|\mu_1,\sigma_1)} \\ &= \frac{1}{\sqrt{2\pi(1-\varrho^2)}\,\sigma_2} \exp\left\{-\frac{[x_2-\mu_2-(\varrho\sigma_2/\sigma_1)(x_1-\mu_1)]^2}{2(1-\varrho^2)\sigma_2{}^2}\right\}, \end{aligned} \tag{1-13}$$

which is a univariate normal density function with (conditional) mean

$$\mathrm{E}\,\{x_2|x_1\} = \mu_2 + \frac{\varrho\sigma_2}{\sigma_1}(x_1 - \mu_1) \tag{1-14}$$

and variance

$$\mathrm{var}\,\{x_2|x_1\} = (1 - \varrho^2)\sigma_2{}^2. \tag{1-15}$$

**Problem 1-8.** Show that, if $x_1$ and $x_2$ are jointly normal with means $\mu_1$ and $\mu_2$, variances $\sigma_1{}^2$ and $\sigma_2{}^2$, respectively, and correlation coefficient $\varrho$, then for a given $x_1$, the best estimate $v$ of $x_2$ is the linear function of $x_1$ (1-14)—best in the sense that it minimizes the expectation of any nondecreasing function of the error $|x_2 - v(x_1)|$, such as the mean-squared error $\mathrm{E}\,\{(x_2 - v)^2\}$.

**Correlation and Independence.** When $\varrho = 0$, (1-13) does not depend on $x_1$; that is, information about the value of $x_1$ is useless in predicting the value of $x_2$. In this case, (1-12) becomes

$$p(x_1,x_2|\mu_1,\mu_2,0,\sigma_1,\sigma_2) = \frac{1}{2\pi\sigma_1\sigma_2} \exp\left[ -\frac{(x_1 - \mu_1)^2}{2\sigma_1{}^2} - \frac{(x_2 - \mu_2)^2}{2\sigma_2{}^2} \right], \tag{1-16}$$

which is simply the product of two density functions of the form (1-3). Thus, for two jointly normal random variables, i.e., for two random variables with a normal bivariate distribution, the vanishing of their correlation coefficient implies their statistical independence.

If their joint density function is not of the form (1-12), then $\varrho = 0$ does *not* imply statistical independence. For example, if $x_1$ is a zero-mean normal random variable, and if $x_2$ takes each of the values $\pm x_1$ with probability $\frac{1}{2}$, then $x_2$ too is a normal, zero-mean random variable. Since for any $x_1$ the expectation of $x_2$ is zero, we have $\mathrm{E}\,\{x_1x_2\} = 0$, and so their correlation coefficient vanishes. But $x_1$ and $x_2$ do not have a joint density function of the form (1-12), and they are not statistically independent, for they always have exactly the same absolute value, and a knowledge of one makes possible the precise prediction of the other except for sign.

On the other hand, independent random variables $x_1$ and $x_2$ are *always* uncorrelated, for in this case

$$\varrho = \frac{\mathrm{E}\,\{(x_1 - \mu_1)(x_2 - \mu_2)\}}{\sigma_1\sigma_2} = \frac{\mathrm{E}\,\{x_1 - \mu_1\} \cdot \mathrm{E}\,\{x_2 - \mu_2\}}{\sigma_1\sigma_2} = 0 \cdot 0 = 0$$

regardless of whether $x_1$ and $x_2$ are normal.

**Problem 1-9.** Show that, if the coordinate axes of Fig. 1-3 are rotated until they are parallel to the axes of the ellipses, the correlation coefficient of the rotated coordinates $x_1'$ and $x_2'$ will be zero, and their standard deviations will be the sum and difference of the two quantities

$$\tfrac{1}{2}\sqrt{\sigma_1{}^2 + \sigma_2{}^2 \pm 2\sqrt{1 - \varrho^2}\,\sigma_1\sigma_2}.$$

In this way any jointly normal $x_1$ and $x_2$ can be expressed in terms of two statistically independent normal random variables as $x_1 = x_1' \cos\alpha - x_2' \sin\alpha$ and $x_2 = x_1' \sin\alpha + x_2' \cos\alpha$. (They can also be expressed as linear functions of two independent normal random variables in many other ways, e.g., as

(1-13) suggests, by putting $x_2 = cx_1 + x_0$ with $c$ a constant and $x_0$ normal and independent of $x_1$.)

**Circular Normal and Rayleigh Distributions.** When $\varrho = 0$, the axes of the ellipses in Fig. 1-3 are parallel to the $x_1$ and $x_2$ axes, and when, in addition, $\sigma_1 = \sigma_2 = \sigma$, the ellipses become circles, and the density function becomes circularly symmetric about the point $(\mu_1,\mu_2)$. In fact, $p(x_1,x_2|\mu_1,\mu_2,0,\sigma,\sigma)$ is the surface obtained by rotating Fig. 1-1 about its axis of symmetry. With $\mu_1 = \mu_2 = 0$ in addition, the probability that the point $(x_1,x_2)$ falls within a small area $dx_1\,dx_2$ becomes

$$p(x_1,x_2|0,0,0,\sigma,\sigma)\,dx_1\,dx_2 = \frac{1}{2\pi\sigma^2}\exp\left(-\frac{x_1^2 + x_2^2}{2\sigma^2}\right)dx_1\,dx_2. \tag{1-17}$$

Changing to polar coordinates $(a,\phi)$ with

$$x_1 = a\cos\phi \qquad \text{and} \qquad x_2 = a\sin\phi,$$

we can write this probability as

$$\frac{1}{2\pi\sigma^2}\exp\left(-\frac{a^2}{2\sigma^2}\right)a\,da\,d\phi, \tag{1-18}$$

since in polar coordinates an incremental area is expressed as $a\,da\,d\phi$. The factor $a$ here is the jacobian

$$J\left(\frac{x_1,x_2}{a,\phi}\right) = \begin{vmatrix} \dfrac{\partial x_1}{\partial a} & \dfrac{\partial x_2}{\partial a} \\ \dfrac{\partial x_1}{\partial \phi} & \dfrac{\partial x_2}{\partial \phi} \end{vmatrix} = a. \tag{1-19}$$

Now (1-18) can be factored as the product of two probabilities,

$$\frac{d\phi}{2\pi}\cdot\frac{a\,da}{\sigma^2}\exp\left(-\frac{a^2}{2\sigma^2}\right),$$

showing that $\phi$ is uniformly distributed (i.e., has a constant probability density) between 0 and $2\pi$ and that $a$ is statistically independent of $\phi$ and is distributed between 0 and $\infty$ with the probability density function

$$p(a|\sigma) = \frac{a}{\sigma^2}e^{-a^2/2\sigma^2} \tag{1-20}$$

(Fig. 1-4), which is known as the Rayleigh density function. The *mode* of this distribution, where (1-20) is maximum, occurs at $a = \sigma$, and its second moment is $\text{E}\{a^2\} = \text{E}\{x_1^2\} + \text{E}\{x_2^2\} = 2\sigma^2$.

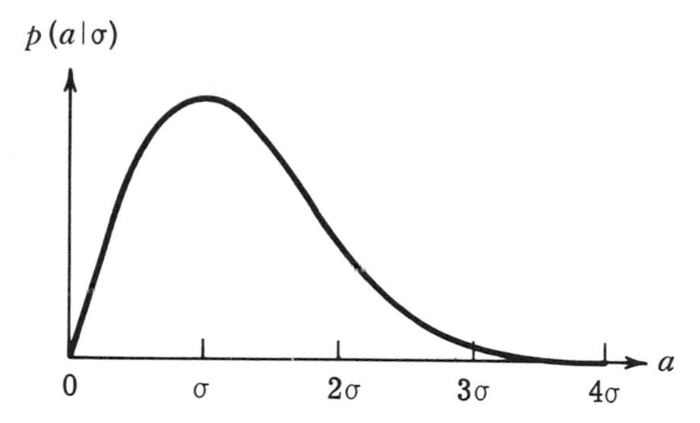


***Figure 1-4.*** The Rayleigh probability density function for $a \geq 0$. For $a \leq 0$ it is zero.

**Problem 1-10.** Show that the $k$th moment of the Rayleigh distribution is $\mathrm{E}\{a^k\} = 2 \cdot 4 \cdot 6 \cdots k\sigma^k$ when $k$ is even. Also show that when $k$ is odd $\mathrm{E}\{a^k\} = 1 \cdot 3 \cdot 5 \cdots k\sqrt{\tfrac{1}{2}\pi}\,\sigma^k$, and express both of these results in a single formula.

**Problem 1-11.** For $x_1 = a \cos \phi$ and $x_2 = a \sin \phi$ with $a$ constant and $\phi$ uniformly distributed between 0 and $2\pi$, find the correlation coefficient of $x_1$ and $x_2$, and determine whether they are statistically independent.

**Problem 1-12.** Repeat Prob. 1-11 with $a$ this time Rayleigh-distributed and statistically independent of $\phi$.

**Matrix Notation.** In preparation for the multivariate normal distribution, we shall now express the bivariate normal density function (1-12) in matrix notation. Thus we let $M_{11} = \sigma_1{}^2$ and $M_{22} = \sigma_2{}^2$ denote the variances of $x_1$ and $x_2$, and we let

$$M_{12} = M_{21} = \varrho\sigma_1\sigma_2 = \mathrm{E}\{(x_1 - \mu_1)(x_2 - \mu_2)\}$$

denote their *covariance*. Hence $M_{jk} = \mathrm{E}\{(x_j - \mu_j)(x_k - \mu_k)\}$. In terms of these quantities, (1-12) takes the form

$$\frac{\exp\left[-\dfrac{M_{22}(x_1-\mu_1)^2 - (M_{12}+M_{21})(x_1-\mu_1)(x_2-\mu_2) + M_{11}(x_2-\mu_2)^2}{2(M_{11}M_{22} - M_{12}M_{21})}\right]}{2\pi\sqrt{M_{11}M_{22} - M_{12}M_{21}}}. \tag{1-21}$$

If we denote by $\mathbf{M}$ the *covariance matrix*

$$\mathbf{M} = \begin{Vmatrix} M_{11} & M_{12} \\ M_{21} & M_{22} \end{Vmatrix},$$

with determinant

$$|\mathbf{M}| = M_{11}M_{22} - M_{12}M_{21}$$

and inverse

$$\mathbf{M}^{-1} = \frac{1}{|\mathbf{M}|}\begin{Vmatrix} M_{22} & -M_{21} \\ -M_{12} & M_{11} \end{Vmatrix}$$

and by $\mathbf{x}$ and $\boldsymbol{\mu}$ the single-column matrices or "vectors"

$$\mathbf{x} = \begin{Vmatrix} x_1 \\ x_2 \end{Vmatrix} \qquad \text{and} \qquad \boldsymbol{\mu} = \begin{Vmatrix} \mu_1 \\ \mu_2 \end{Vmatrix}$$

whose transposes are the row vectors

$$\tilde{\mathbf{x}} = \|x_1 \quad x_2\| \qquad \text{and} \qquad \tilde{\boldsymbol{\mu}} = \|\mu_1 \quad \mu_2\|,$$

respectively, then the bivariate normal probability density function (1-21) can be written simply as

$$p(\mathbf{x}|\boldsymbol{\mu},\mathbf{M}) = \frac{1}{(2\pi)^{m/2}\sqrt{|\mathbf{M}|}} \exp\left[-\tfrac{1}{2}(\tilde{\mathbf{x}} - \tilde{\boldsymbol{\mu}})\mathbf{M}^{-1}(\mathbf{x} - \boldsymbol{\mu})\right], \qquad (1\text{-}22)$$

with $m$, the dimensionality of the vectors, equal to 2. Notice that the mean vector and the covariance matrix can be expressed quite simply as

$$\boldsymbol{\mu} = \mathrm{E}\,\{\mathbf{x}\} \qquad \text{and} \qquad \mathbf{M} = \mathrm{E}\,\{(\mathbf{x} - \boldsymbol{\mu})(\tilde{\mathbf{x}} - \tilde{\boldsymbol{\mu}})\}, \qquad (1\text{-}23)$$

respectively.

## □ THE MULTIVARIATE NORMAL DISTRIBUTION

In its multivariate form, the central limit theorem of probability theory[1] states that, if an $m$-dimensional vector $\mathbf{x}$ is the sum of a great many statistically independent $m$-dimensional vectors whose components all have variances small compared to those of their respective sums, then the probability density function of $\mathbf{x}$ tends toward the multivariate normal form described by (1-22), $\boldsymbol{\mu}$ being the sum of the means of the summands and $\mathbf{M}$ the sum of their covariance matrices. For $m = 1$ and $m = 2$, the multivariate normal density function (1-22) becomes (1-3) and (1-21), respectively.

The mean vector $\boldsymbol{\mu}$ can take any value at all, but the covariance matrix $\mathbf{M}$ cannot, since for any $m$-dimensional vector $\mathbf{z}$ with real com-

[1] Cramér, *ibid.*, chap. 24, pp. 310–317, The Normal Distribution; John M. Wozencraft and Irwin M. Jacobs, "Principles of Communication Engineering," sec. 3.3, pp. 156–170, The Multivariate Central Limit Theorem, John Wiley & Sons, Inc., New York, 1965.

ponents we have

$$\begin{aligned}\tilde{\mathbf{z}}\mathbf{M}\mathbf{z} &= \tilde{\mathbf{z}}\,\mathrm{E}\,\{(\mathbf{x}-\boldsymbol{\mu})(\tilde{\mathbf{x}}-\tilde{\boldsymbol{\mu}})\}\mathbf{z} \\ &= \mathrm{E}\,\{[\tilde{\mathbf{z}}(\mathbf{x}-\boldsymbol{\mu})][(\tilde{\mathbf{x}}-\tilde{\boldsymbol{\mu}})\mathbf{z}]\} \\ &= \mathrm{E}\,\{[(\tilde{\mathbf{x}}-\tilde{\boldsymbol{\mu}})\mathbf{z}]^2\} \\ &\geq 0, \end{aligned} \tag{1-24}$$

whether $\mathbf{x}$ is normally distributed or not. A matrix $\mathbf{M}$ with the property (1-24) for all real $\mathbf{z}$ is called *positive semidefinite*, and any such matrix can be the covariance matrix of a multivariate normal distribution.

If $\mathbf{M}$ has an inverse $\mathbf{M}^{-1}$, and we set $\mathbf{z} = \mathbf{M}^{-1}(\mathbf{x}-\boldsymbol{\mu})$ in (1-24), we find that $(\tilde{\mathbf{x}}-\tilde{\boldsymbol{\mu}})\mathbf{M}^{-1}(\mathbf{x}-\boldsymbol{\mu}) \geq 0$. Hence the contours of constant normal probability density (1-22) are similar, coaxial ellipsoids with center $\boldsymbol{\mu}$. By rotating the coordinate frame so that its axes are parallel to those of the ellipsoids, the integral of (1-22) over the entire $m$-dimensional space of the vector $\mathbf{x}$ can be factored as the product of $m$ integrals of univariate normal density functions (1-3), since cross-product terms in the exponent then disappear, and the integral of (1-22) can thus be shown to be unity. (See Problem 1-9.)

When $|\mathbf{M}| = 0$, and $\mathbf{M}^{-1}$ therefore does not exist, the distribution is called *singular*. In this case at least one of the semiaxes of the ellipsoids is zero, and $\mathbf{x}$ is confined to a linear manifold of lower dimensionality, for there exist one or more linear relationships among its components. The elimination of the dependent components permits (1-22) to be used, with $m$ equal to the rank of $\mathbf{M}$, which is the number of nonzero semiaxes of the ellipsoids.

If $\mathbf{x}$ is normally distributed, then any subset of its components is also normally distributed, for its probability density function is again of the form (1-22), as we can see by completing the square in any omitted components and integrating. Likewise, if the components of $\mathbf{y}$ are linear functions of the components of $\mathbf{x}$; that is, if $\mathbf{y} = \mathbf{K}\mathbf{x} + \mathbf{k}$ with $\mathbf{K}$ a constant matrix and $\mathbf{k}$ a constant vector, then $\mathbf{y}$ too is normally distributed, its mean being $\mathrm{E}\,\{\mathbf{y}\} = \mathbf{K}\boldsymbol{\mu} + \mathbf{k}$ and its covariance matrix

$$\mathrm{E}\,\{(\mathbf{y}-\mathrm{E}\,\mathbf{y})(\tilde{\mathbf{y}}-\mathrm{E}\,\tilde{\mathbf{y}})\} = \mathrm{E}\,\{\mathbf{K}(\mathbf{x}-\boldsymbol{\mu})(\tilde{\mathbf{x}}-\tilde{\boldsymbol{\mu}})\tilde{\mathbf{K}}\} = \mathbf{K}\mathbf{M}\tilde{\mathbf{K}}.$$

**Problem 1-13.** Verify all of the foregoing matrix manipulations by writing them out in terms of summations of products of components.

**Characteristic Function.** The characteristic function of the distribution of $x_1, x_2, \ldots, x_m$ is defined as the expectation of

$$\exp\,[i(z_1x_1 + z_2x_2 + \cdots + z_mx_m)] = e^{i\tilde{\mathbf{z}}\mathbf{x}}, \tag{1-25}$$

where $i^2 = -1$ and $z_1, z_2, \ldots, z_m$ are the arguments of the characteristic function. Multiplying (1-25) by (1-22) and integrating over all $\mathbf{x}$, we obtain the characteristic function of the multivariate normal distribution,

$$\mathrm{E}\,\{e^{i\tilde{\mathbf{z}}\mathbf{x}}\} = \frac{1}{(2\pi)^{m/2}\sqrt{|\mathbf{M}|}} \int \exp\,[i\tilde{\mathbf{z}}\mathbf{x} - \tfrac{1}{2}(\tilde{\mathbf{x}} - \tilde{\boldsymbol{\mu}})\mathbf{M}^{-1}(\mathbf{x} - \boldsymbol{\mu})]\,d\mathbf{x},$$

where $d\mathbf{x} = dx_1\,dx_2 \cdots dx_m$, and the limits of integration are $-\infty$ and $\infty$ on each component of $\mathbf{x}$. If $\mathbf{y} + \boldsymbol{\mu} + i\mathbf{M}\mathbf{z}$ is substituted for $\mathbf{x}$, with $d\mathbf{y} = d\mathbf{x}$, this expression becomes

$$\begin{aligned}\mathrm{E}\,\{e^{i\tilde{\mathbf{z}}\mathbf{x}}\} &= \frac{\exp\,(i\tilde{\mathbf{z}}\boldsymbol{\mu} - \tfrac{1}{2}\tilde{\mathbf{z}}\mathbf{M}\mathbf{z})}{(2\pi)^{m/2}\sqrt{|\mathbf{M}|}} \int \exp\,(-\tfrac{1}{2}\tilde{\mathbf{y}}\mathbf{M}^{-1}\mathbf{y})\,d\mathbf{y} \\ &= \exp\,(i\tilde{\mathbf{z}}\boldsymbol{\mu} - \tfrac{1}{2}\tilde{\mathbf{z}}\mathbf{M}\mathbf{z}),\end{aligned} \tag{1-26}$$

since the substitution of $\mathbf{x} - \boldsymbol{\mu}$ for $\mathbf{y}$ yields the integral of (1-22) over all $\mathbf{x}$, which must be unity.

In the univariate case, $\mathbf{x}$, $\boldsymbol{\mu}$, $\mathbf{M}$, and $\mathbf{z}$ have only a single component, $x$, $\mu$, $\sigma^2$, and $z$, respectively, and (1-26) becomes

$$\mathrm{E}\,\{e^{izx}\} = \int_{-\infty}^{\infty} e^{izx}p(x)\,dx = \exp\,(iz\mu - \tfrac{1}{2}\sigma^2 z^2). \tag{1-27}$$

By its definition the characteristic function of a single random variable is the Fourier transform of the probability density function, and indeed, (1-27) is the Fourier transform of (1-3). In the multivariate case it is the multidimensional Fourier transform, and the characteristic function, for example, (1-26), therefore determines the probability density function, for example, (1-22), through inverse Fourier transformation.

**The Sum of Independent Random Vectors.** If $\mathbf{x}$ and $\mathbf{x}'$ are independent $m$-component random vectors, the characteristic function of their sum $\mathbf{x}'' = \mathbf{x} + \mathbf{x}'$ is $\mathrm{E}\,\{e^{i\tilde{\mathbf{z}}\mathbf{x}''}\} = \mathrm{E}\,\{e^{i\tilde{\mathbf{z}}\mathbf{x}}\}\,\mathrm{E}\,\{e^{i\tilde{\mathbf{z}}\mathbf{x}'}\}$, the product of their separate characteristic functions. Thus, if $\mathbf{x}$ and $\mathbf{x}'$ are normal, with characteristic functions $\exp\,(i\tilde{\mathbf{z}}\boldsymbol{\mu} - \tfrac{1}{2}\tilde{\mathbf{z}}\mathbf{M}\mathbf{z})$ and $\exp\,(i\tilde{\mathbf{z}}\boldsymbol{\mu}' - \tfrac{1}{2}\tilde{\mathbf{z}}\mathbf{M}'\mathbf{z})$, respectively, that of $\mathbf{x}''$ is $\exp\,[i\tilde{\mathbf{z}}(\boldsymbol{\mu} + \boldsymbol{\mu}') - \tfrac{1}{2}\tilde{\mathbf{z}}(\mathbf{M} + \mathbf{M}')\mathbf{z}]$. Hence $\mathbf{x}''$ too is normal, with mean $\boldsymbol{\mu} + \boldsymbol{\mu}'$ and covariance matrix $\mathbf{M} + \mathbf{M}'$; that is, the convolution $\int p(\mathbf{x})q(\mathbf{x}'' - \mathbf{x})\,d\mathbf{x}$ of two normal density functions $p(\mathbf{x})$ and $q(\mathbf{x}')$ is another normal density function $r(\mathbf{x}'')$.

**Moments.** Writing

$$\begin{aligned}\mathrm{E}\,\{e^{i\tilde{\mathbf{z}}\mathbf{x}}\} &= 1 + \mathrm{E}\,\{i\tilde{\mathbf{z}}\mathbf{x}\} + \tfrac{1}{2}\,\mathrm{E}\,\{(i\tilde{\mathbf{z}}\mathbf{x})^2\} + \cdots \\ &= 1 + i\tilde{\mathbf{z}}\,\mathrm{E}\,\{\mathbf{x}\} - \tfrac{1}{2}\,\mathrm{E}\,\{\tilde{\mathbf{z}}\mathbf{x}\tilde{\mathbf{x}}\mathbf{z}\} + \cdots \\ &= 1 + i\tilde{\mathbf{z}}\,\mathrm{E}\,\{\mathbf{x}\} - \tfrac{1}{2}\tilde{\mathbf{z}}\,\mathrm{E}\,\{\mathbf{x}\tilde{\mathbf{x}}\}\mathbf{z} + \cdots\end{aligned}$$

and comparing it with (1-26), which can likewise be expanded as

$$\begin{aligned}\mathrm{E}\,\{e^{i\tilde{\mathbf{z}}\mathbf{x}}\} &= 1 + i\tilde{\mathbf{z}}\boldsymbol{\mu} - \tfrac{1}{2}\tilde{\mathbf{z}}\mathbf{M}\mathbf{z} - \tfrac{1}{2}(\tilde{\mathbf{z}}\boldsymbol{\mu})^2 + \cdots \\ &= 1 + i\tilde{\mathbf{z}}\boldsymbol{\mu} - \tfrac{1}{2}\tilde{\mathbf{z}}(\mathbf{M} + \boldsymbol{\mu}\tilde{\boldsymbol{\mu}})\mathbf{z} + \cdots,\end{aligned}$$

we see, by equating the coefficients of terms in $z_j$ and $z_jz_k$, respectively, that $\mathrm{E}\{\mathbf{x}\} = \boldsymbol{\mu}$ and that $\mathrm{E}\{\mathbf{x}\tilde{\mathbf{x}}\} = \mathbf{M} + \boldsymbol{\mu}\tilde{\boldsymbol{\mu}}$; that is,

$$\mathrm{E}\{(\mathbf{x} - \boldsymbol{\mu})(\tilde{\mathbf{x}} - \tilde{\boldsymbol{\mu}})\} = \mathbf{M}$$

for the distribution whose characteristic function is (1-26) and whose density function is (1-22). Thus we have verified that the $\boldsymbol{\mu}$ and $\mathbf{M}$ in (1-22) are indeed the mean and covariance of the distribution that it describes.

We can also use (1-26) in a similar manner to find the higher moments of the multivariate normal distribution. It will suffice to treat the case where $\boldsymbol{\mu} = \mathbf{0}$, that is, where all $x_j$ have mean value zero, and the moment $\mathrm{E}\{x_1x_2 \cdots x_m\}$ will be sufficiently general. Thus, in the series

$$\mathrm{E}\{e^{i\tilde{\mathbf{z}}\mathbf{x}}\} = \sum_{s=0}^{\infty} \frac{i^s}{s!} \mathrm{E}\{(\tilde{\mathbf{z}}\mathbf{x})^s\}$$

$\mathrm{E}\{x_1x_2 \cdots x_m\}$ appears only in the term $(i^m/m!)\,\mathrm{E}\{(\tilde{\mathbf{z}}\mathbf{x})^m\}$ for which $s = m$, and there it appears $m!$ times; consequently its coefficient is $i^m z_1z_2 \cdots z_m$. Hence $\mathrm{E}\{x_1x_2 \cdots x_m\}$ is $i^{-m}$ times the coefficient of $z_1z_2 \cdots z_m$ in the power-series expansion of (1-26), which is here $\exp(-\frac{1}{2}\tilde{\mathbf{z}}\mathbf{M}\mathbf{z})$.

If $m$ is odd, no such term appears, that is, $\mathrm{E}\{x_1x_2 \cdots x_m\} = 0$. If $m$ is even, only the term

$$\frac{(-\frac{1}{2})^{m/2}}{(\frac{1}{2}m)!}\left(\sum_{j,k=1}^{m} M_{jk}z_jz_k\right)^{m/2}$$

in the power series contains such a term, and in it the coefficient of $z_1z_2 \cdots z_m$ is $(-\frac{1}{2})^{m/2}/(\frac{1}{2}m)!$ times the summation, over all permutations of the subscripts, of $M_{12}M_{34} \cdots M_{(m-1)m}$. Since permuting these $\frac{1}{2}m$ factors and interchanging the two subscripts on any factor does not alter their product, for each different way of dividing the $m$ subscripts into $\frac{1}{2}m$ pairs there are $2^{m/2}(\frac{1}{2}m)!$ equal terms. Thus the coefficient of $z_1z_2 \cdots z_m$ is $(-1)^{m/2}$ times the summation of the product $M_{12}M_{34} \cdots M_{(m-1)m}$ over all $2^{-m/2}m!/(\frac{1}{2}m)! = 1 \cdot 3 \cdot 5 \cdots (m-1)$ permutations of the subscripts yielding different values for the product (when account is taken of the equality $M_{jk} = M_{kj}$ and of the commutativity of multiplication). Hence, for even $m$ with $\boldsymbol{\mu} = \mathbf{0}$,

$$\mathrm{E}\{x_1x_2 \cdots x_m\} = \sum_{\substack{\text{All permutations of} \\ \text{subscripts giving} \\ \text{different products}}}^{1\cdot3\cdot5\cdots(m-1)\text{ terms}} M_{12}M_{34} \cdots M_{(m-1)m}. \tag{1-28}$$

For example, $$\mathrm{E}\{x_1x_2\} = M_{12}$$

and $$\mathrm{E}\{x_1x_2x_3x_4\} = M_{12}M_{34} + M_{13}M_{24} + M_{14}M_{23}. \tag{1-29}$$

If $\boldsymbol{\mu} \neq \mathbf{0}$, we can write

$$\mathrm{E}\,\{x_1x_2 \cdots x_m\} = \mathrm{E}\,\{(y_1 + \mu_1)(y_2 + \mu_2) \cdots (y_m + \mu_m)\},$$

where $\mathbf{y} = \mathbf{x} - \boldsymbol{\mu}$ is normally distributed with zero mean. Expanding the right-hand side as a polynomial in the components of $\mathbf{y}$, we can therefore use (1-28) to find the mean value of each term, and the sum of these mean values is the desired moment $\mathrm{E}\,\{x_1x_2 \cdots x_m\}$.

**Problem 1-14.** Derive (1-4) from (1-28) by supposing that $\mathbf{x}$ is normal with all of its components equal.

**Problem 1-15.** Show that, for $x_1$ and $x_2$ obeying the bivariate normal distribution with zero means, $\mathrm{E}\,\{x_1^2x_2^2\} = M_{11}M_{22} + 2M_{12}^2 = (1 + 2\varrho^2)\sigma_1^2\sigma_2^2$.

**Problem 1-16.** Show that, if $x_1$, $x_2$, $x_3$ are independent, zero-mean, normal random variables, then $y_1 = x_2 \operatorname{sgn} x_3$, $y_2 = x_3 \operatorname{sgn} x_1$, and $y_3 = x_1 \operatorname{sgn} x_2$ do not obey a trivariate normal distribution although every pair $y_j$, $y_k$ is bivariate normal, sgn $x$ being $\pm 1$ according to the sign of $x$.

# chapter 2 RANDOM PROCESSES

In Chap. 1 we dealt with finite sets of random variables; in this chapter we shall be concerned with random functions, for example, $u(t)$, which involve an infinite set of random variables—$u(t)$ for each real value of $t$. Here $u(t)$ may represent a random signal or noise waveform. A random process, then, is the source of such a random function of time, which we might designate more fully as $u(t,\omega)$. Here $\omega$ is an abstract parameter, selected at random by the process, that determines just which function $u(t)$ will result. Such a random process is fully specified if, for any set of $m$ values ($m = 1, 2, \ldots$) of $t$, say $t_1, t_2, \ldots, t_m$, we know the joint distribution of $x_1 = u(t_1), x_2 = u(t_2), \ldots, x_m = u(t_m)$. These distributions must be implied by an assumed mathematical model of the process. They cannot be determined experimentally without some assumption about the nature of the process (such as ergodicity, which we shall take up later) because, in the interval $-\infty < t < \infty$, a random process can produce only a single $u(t)$.

**Gaussian Random Process.** The gaussian random process furnishes an important example. Here the $m$-variate density functions all have the normal form (1-22) with $\mu_j = \mathrm{E}\,\{u(t_j)\} = \mu(t_j)$ given by any prescribed $\mu(t)$ and $M_{jk} = \mathrm{E}\,\{[u(t_j) - \mu(t_j)][u(t_k) - \mu(t_k)]\} = M(t_j,t_k)$ given by any $M(t,t') = M(t',t)$ that yields only positive semidefinite covariance matrices.

**Sinusoidal Gaussian Process.** A particular case of interest is the *sinusoidal* gaussian random process. Here we take the mean to be $\mu(t) = 0$ for all $t$ and the covariance

$$M(t,t') = \sigma^2 \cos 2\pi F(t - t'). \tag{2-1}$$

We shall show that $u(t)$ must consequently be a sinusoid of frequency $F$ by noticing that, for any $t$, $t'$, $t''$,

$$\begin{aligned}
&\mathrm{E}\,\{[u(t)\sin 2\pi F(t'-t'') + u(t')\sin 2\pi F(t''-t) + u(t'')\sin 2\pi F(t-t')]^2\} \\
&\quad = \sigma^2 \sin^2 2\pi F(t'-t'') + \sigma^2 \sin^2 2\pi F(t''-t) + \sigma^2 \sin^2 2\pi F(t-t') \\
&\qquad + 2\sigma^2 \cos 2\pi F(t-t') \sin 2\pi F(t'-t'') \sin 2\pi F(t''-t) \\
&\qquad + 2\sigma^2 \cos 2\pi F(t'-t'') \sin 2\pi F(t''-t) \sin 2\pi F(t-t') \\
&\qquad + 2\sigma^2 \cos 2\pi F(t''-t) \sin 2\pi F(t-t') \sin 2\pi F(t'-t'') \\
&\quad = 0.
\end{aligned} \tag{2-2}$$

Since the square of the quantity in brackets cannot be negative, it follows that, with unit probability, it is zero. Hence, for

$$\sin 2\pi F(t''-t') \neq 0,$$

we have

$$u(t) = \frac{u(t'')\sin 2\pi F(t-t') - u(t')\sin 2\pi F(t-t'')}{\sin 2\pi F(t''-t')}, \tag{2-3}$$

that is, $u(t)$ is the sum of two sinusoids of frequency $F$ and is determined for all $t$ by its values at two instants. It follows that its covariance matrices for $m \geq 2$ all have rank 2 and that $u(t)$ has the form

$$u(t) = a \cos (2\pi Ft + \phi), \tag{2-4}$$

with $a$ and $\phi$ random constants.

Putting $t' = 0$ and $t'' = 1/4F$, we see from (2-1) that $u(t')$ and $u(t'')$ are then independent zero-mean normal random variables of variance $\sigma^2$. They therefore obey the circular bivariate normal distribution, and it follows by (1-18) that the amplitude $a$ of $u(t)$ is Rayleigh-distributed and that its phase $\phi$ is statistically independent of $a$ and is uniformly distributed between 0 and $2\pi$.

A gaussian random process is often called *gaussian noise,* and this sinusoidal gaussian random process may be thought of as the special case in which the noise has been passed through a filter of zero bandwidth which passes only the frequency $F$. Accordingly, its output must be a sinusoid (2-4) of this frequency.

**Problem 2-1.** Verify that (2-4), with $a$ Rayleigh-distributed and $\phi$ independently uniformly distributed, has mean zero and covariance function (2-1), and show that, for any $t$ and $t'$, $u(t)$ and $u(t')$ obey a bivariate normal distribution with these parameters.

**Problem 2-2.** Derive (2-3) from (2-4), thus showing the motivation for (2-2).

**Problem 2-3.** Describe the case of the sinusoidal gaussian process with $F = 0$; it may be called a *constant gaussian process.*

**Random Telegraph Signals.** The notion of a random process is quite general and includes such things as a person's speech waveform and the current through a telegraph key. The latter can easily be made to furnish useful examples by supposing appropriate statistics for the motion of the key, and we shall consider two particular kinds, called *asynchronous* and *synchronous*. Both kinds of random telegraph signals[1] will take only the values $u(t) = \pm 1$, representing the two positions of the telegraph key, with occasional jumps from 1 to $-1$ or from $-1$ to 1.

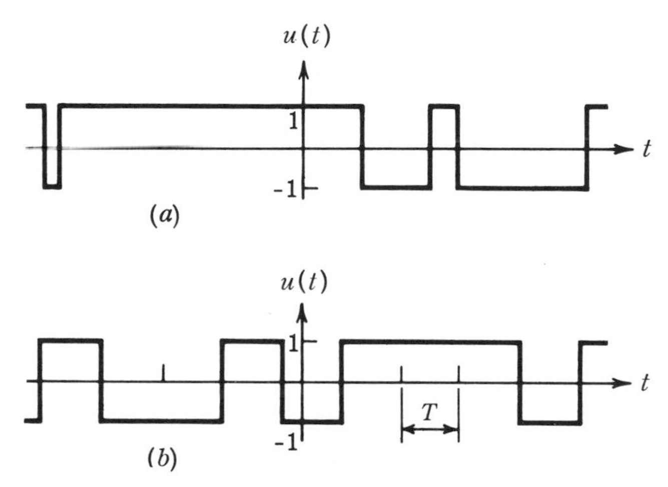


*Figure 2-1.* Random telegraph signals: (*a*) asynchronous and (*b*) synchronous.

In the asynchronous case, we suppose that the times of occurrence of these jumps form a *Poisson process*, in which the probability that a jump occurs between $t$ and $t + dt$ is $\nu\, dt$, independently of whatever occurs outside this interval, where $\nu$ is a constant.

In the case of the synchronous random telegraph signal, we assume that the time scale is divided into equal intervals of length $T$, and that $u(t)$ is constant during each interval, taking either of the values $\pm 1$ with probability $\frac{1}{2}$ independently of its values in other intervals (see Fig. 2-1).

## ☐ STATIONARITY

A random process is called *stationary* if all of its $m$-variate probability distributions ($m = 1, 2, \ldots$) depend only on time differences, such as $t_2 - t_1$, $t_3 - t_1$, and so on, and not on the time itself, for example, $t_1$. A

[1] S. O. Rice, Mathematical Analysis of Random Noise, *Bell System Tech. J.*, **23**:282–332 (1944) and **24**:46–157 (1945); reprinted in Nelson Wax (ed.), "Noise and Stochastic Processes," pp. 133–294, Dover Publications, Inc., New York, 1954.

stationary random process can be defined equivalently as one whose probability distributions are all invariant under every time translation, i.e., replacement of $t$ by $t - t_0$. Hence a sample (i.e., a record) of a stationary random process gives no clue to the time of occurrence of any of its features.

Stationarity does not imply that the sample itself is somehow the same at all times, but clearly the univariate distribution of $u(t)$ must be independent of $t$, and hence its mean, variance, and all higher moments (if they exist)[1] must be constant.

From the definition it follows that a gaussian random process is stationary if and only if its mean is a constant, that is, $\mu(t) = \mu$, and its covariance is a function only of the difference of its two arguments, that is, $M(t,t') = M(t - t')$, since these two functions determine all of its distributions. The sinusoidal gaussian process is evidently of this form and is therefore stationary.

In general, however, a sinusoidal random process

$$u(t) = a \cos (2\pi F t + \phi)$$

with the constants $a$ and $\phi$ arbitrarily distributed will not be stationary, since it is more likely to be zero near some times than near others, and its univariate distribution will therefore be time-dependent.

If started at some time with probability $\frac{1}{2}$ for either of the values $u(t) = \pm 1$ and extended toward both $t = \pm \infty$ from there, the asynchronous random telegraph signal will be stationary, for its behavior is not tied to any particular origin of time. To make the synchronous random telegraph signal stationary, we must start it similarly *and* distribute the epoch of its jumps uniformly over an interval of length $T$ so that the jumps will give no clue to the time $t$.

**Problem 2-4.** (*a*) Is the sum of two independent stationary random processes stationary? (*b*) The product? (*c*) Is the derivative of a stationary random process stationary? (*d*) Show that the integral of a stationary random process is not necessarily stationary.

## ☐ ERGODICITY

An *ensemble* may be defined as a large set of randomly selected sample functions $u(t)$ that might be generated by a random process,[2] that is,

[1] For example, the distribution with density function $p(x) = 1/(x \ln^2 x)$ for $x > e$ and $p(x) = 0$ for $x \leq e$ has no moments at all of positive order.

[2] More precisely, an ensemble is the (not necessarily large) set of *all* possible samples, together with a probability measure over this set. Our loose definition, however, avoids certain conceptual difficulties.

$u(t,\omega)$ for a large set of random values of the abstract parameter $\omega$. The $m$-variate probability density function of the process is sometimes called an *ensemble statistic* because, when multiplied by $dx_1\,dx_2 \cdots dx_m$, it equals the proportion of the members of the ensemble for which

$$x_j \leq u(t_j) < x_j + dx_j, \qquad \text{with } j = 1, 2, \ldots, m.$$

Thus ensemble statistics are averages with respect to the parameter $\omega$ and are functions of $t_1, t_2, \ldots, t_m$.

From a single sample function $u(t)$ generated by the random process we can form a large, randomly selected set of its translates $u(t - t_0)$ with the random variable $t_0$ distributed uniformly over a long interval $T$

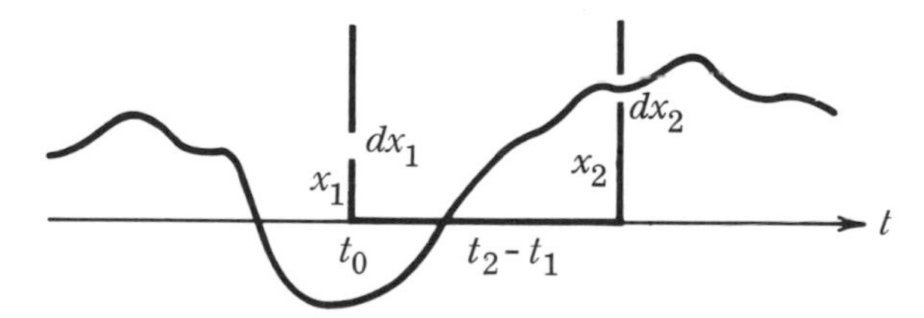


**Figure 2-2.** If a three-sided frame of width $t_2 - t_1$ with windows of width $dx_1$ and $dx_2$ at height $x_1$ and $x_2$ in its left and right sides, respectively, is slid past $u(t)$, the fraction of the time $u(t)$ passes through both windows simultaneously is the time-average $p(x_1,x_2,t_2 - t_1)dx_1dx_2$, which gives the bivariate probability density function of $u(t)$.

and playing a role similar to that of $\omega$ in an ensemble. If, as $T \to \infty$, the proportion of members of this set for which

$$x_j \leq u(t_j - t_0) < x_j + dx_j \qquad \text{with } j = 1, 2, \ldots, m$$

approaches a limit, the quotient of this limit divided by $dx_1\,dx_2 \cdots dx_m$ is the $m$-variate density function of the translates of the given $u(t)$; we refer to such quantities as *time statistics* of the particular $u(t)$ from which they are derived (Fig. 2-2). Because of the way they are obtained, time statistics are always stationary, for they can depend only on time differences $t_j - t_k$, but in general they also depend on $\omega$, that is, they are different for different samples $u(t)$.

However, for some random processes the time statistics of all samples (except a set of total probability zero) are the same. If, in addition, such a random process is stationary, its ensemble statistics will be unaffected by replacing $t$ by $t - t_0$ or by averaging the result over a distribution of $t_0$ that is uniform over an interval whose length is allowed to become infinite. The ensemble statistics are, of course, an average over a large number of random samples, and this time average of the ensemble statistics can, by reversing the order of averaging, be seen to be the ensemble average of the time statistics of the sample functions. With the latter all identical, we thus find that the time statistics of any sample are the same as the ensemble statistics of the random process. In this case, every sample is typical of the entire ensemble, and the random process is called *ergodic.*

The sinusoidal gaussian random process, though stationary, is not ergodic, for different samples will have different constant amplitudes and, therefore, different univariate time statistics; no sample exhibits the Rayleigh distribution of amplitudes characterizing the ensemble.

On the other hand, every stationary gaussian random process whose covariance function $M(\tau)$ goes to zero as $\tau \to \infty$ is ergodic, since widely separated segments of $u(t)$ are in this case almost statistically independent of one another and might as well be segments of *different* samples of the process. A sufficiently long record $u(t)$, therefore, provides the equivalent of an arbitrarily large number of independent samples, and hence its time statistics are the same as the ensemble statistics of the process.

In fact, $M(\infty) = 0$ is a necessary as well as sufficient condition for a stationary gaussian random process to be ergodic, since otherwise each sample will contain a periodic component of constant amplitude. However, $M(\infty) = 0$ is in general neither a necessary nor a sufficient condition for ergodicity (see Prob. 2-6).

Since widely separated segments of a single sample of an asynchronous random telegraph signal tend to be statistically independent, this process is ergodic for the same reason as a stationary gaussian random process with $M(\infty) = 0$. Similarly, when made stationary by a uniform distribution of its epoch, the synchronous random telegraph signal is ergodic too.

**Problem 2-5.** Is the constant gaussian random process (Problem 2-3) ergodic?

**Problem 2-6.** By considering a random telegraph signal with more than one possible constant amplitude and a sinusoidal random process with a single possible amplitude, show that $M(\infty) = 0$ is neither necessary nor sufficient for a stationary random process to be ergodic.

**Problem 2-7.** By considering two sinusoidal random processes having only a single possible amplitude, show that neither the sum nor the product of two independent ergodic processes is necessarily ergodic.

**Problem 2-8.** Is the derivative of an ergodic random process ergodic? Show that the integral of an ergodic random process is not necessarily ergodic.

**Joint Ergodicity.** Two (ergodic) random processes are said to be *jointly ergodic* if the ensemble statistics of the pair of processes are the same as the time statistics of almost every pair of samples obtained by an obvious generalization of Fig. 2-2. When two random processes are jointly ergodic, therefore, any long record of the two as functions of the same time parameter will exhibit the ensemble statistics of the pair.

If, as we saw in Prob. 2-7, the two processes have commensurable periodicities, this cannot be the case, for their periodic components will remain forever in a fixed phase relationship. However, when two random processes are jointly ergodic, any function of the two will also be ergodic. This will be true of a wide variety of pairs of random processes with which we shall deal, such as a signal and a noise or a modulating waveform and the modulated carrier—whenever each is ergodic and the two have no common periodicity of any sort.

**Problem 2-9.** Are two independent synchronous random telegraph signals having the same value of $T$ jointly ergodic? Two independent asynchronous random telegraph signals having the same value of $\nu$? Two independent ergodic gaussian random processes?

**Problem 2-10.** If $u(t,\omega)$ is ergodic, $u(t,\omega)$ and its translate $u(t + \tau, \omega)$ are *dependent* ergodic random processes. Are they jointly ergodic? Is $\int_t^{t+\tau} u(s)\, ds$ ergodic?

**Problem 2-11.** If, throughout each interval of constancy, a random telegraph signal took a new, independent, zero-mean, unit-variance, normally distributed value, would the random process be ergodic? Gaussian?

**Weak Stationarity and Ergodicity.** For many applications, particularly in the next four chapters, it will be sufficient for our random processes to satisfy considerably weaker conditions than the foregoing (strict) stationarity and ergodicity. In connection with power spectra, it will only be necessary that the mean be constant and that the covariance function depend solely on the difference of its two arguments; it will not matter if other ensemble statistics are not stationary. Likewise, every sample will be sufficiently typical of the ensemble if, with unit probability, its time-average mean and covariance function coincide with those of the ensemble. Strict ergodicity will be necessary only later on, especially in information theory.

Every strictly ergodic random process is, *a fortiori*, weakly ergodic, i.e., ergodic "in the wide sense." Although it should be much easier to satisfy the weak condition, the random processes known to be weakly ergodic usually turn out to be strictly ergodic too, and so the distinction loses its importance. We shall therefore not stress which form of ergodicity is needed; in each instance it will be easy to see which one is required.

# chapter 3 THE POWER SPECTRUM

The second moment E $\{u^2(t)\}$ of a random process $u(t)$ is often referred to as its *average power*, since $u^2$ is the power that a voltage $u$ would dissipate in a unit resistance. In general this average power is a function of $t$, but for a stationary $u(t)$ it is a constant. We shall now make use of the frequency response of an arbitrary fixed linear filter to determine how the average power of an *ergodic* random process $u(t)$ is distributed in frequency, i.e., we shall find its power spectrum.

**Linear Filtering.** The frequency response of a fixed linear filter whose (time) response to a unit impulse $\delta(t)$ is $g(t)$ is the Fourier transform

$$G(f) = \int_{-\infty}^{\infty} g(t)e^{-2\pi i f t}\,dt, \tag{3-1}$$

with

$$g(t) = \int_{-\infty}^{\infty} G(f)e^{2\pi i f t}\,df. \tag{3-2}$$

If $u(t)$ is the input to the filter, its output $v(t)$ is the convolution (Fig. 3-1)

$$\begin{aligned} v(t) &= u(t) \star g(t) \\ &= g(t) \star u(t) \\ &= \int_{-\infty}^{\infty} u(t-\tau)g(\tau)\,d\tau. \end{aligned} \tag{3-3}$$

Here the star denotes the (commutative) operation of convolution described by the integral.

In terms of the Fourier transforms $U(f)$ and $V(f)$ of $u(t)$ and $v(t)$, respectively, (3-3) would be written simply $V(f) = U(f)G(f)$. However, because $u(t)$ is ergodic, (3-3) too is ergodic (see Prob. 2-10), and neither tends to zero as $t$ grows infinite. Hence, the integrals defining $U(f)$ and $V(f)$ do not converge, and these transforms do not exist.

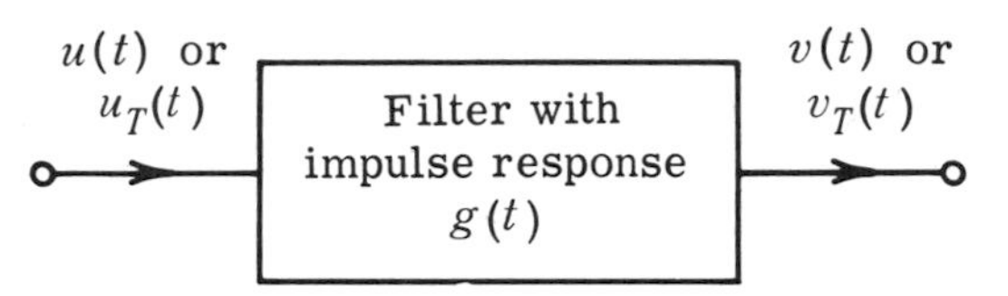


*Figure 3-1.* Filtering an ergodic random process to determine its power spectrum.

To circumvent this difficulty, we truncate $u(t)$ at $t = \pm\frac{1}{2}T$, which we shall later allow to become infinite, and we use the resulting

$$u_T(t) = \begin{cases} u(t) & \text{for } |t| < \frac{1}{2}T, \\ 0 & \text{for } |t| \geq \frac{1}{2}T, \end{cases} \tag{3-4}$$

as the filter input. Consequently, the input and the output (which we do not truncate) have Fourier transforms, say $U_T(f)$ and $V_T(f)$, respectively, with $V_T(f) = U_T(f)G(f)$.

If we multiply

$$U_T(f) = \int_{-\frac{1}{2}T}^{\frac{1}{2}T} u(\tau)e^{-2\pi if\tau}\,d\tau \tag{3-5}$$

by $e^{2\pi ift}$ and take the real part, we get (Fig. 3-2)

$$\text{Re}\,\{U_T(f)e^{2\pi ift}\} = \int_{-\frac{1}{2}T}^{\frac{1}{2}T} u(\tau)\cos 2\pi f(t-\tau)\,d\tau,$$

which is the response of an $LC$ circuit resonant to frequency $f$, with impulse response $\cos 2\pi ft$ for $t > 0$ and $0$ for $t \leq 0$ after having had $u(t)$

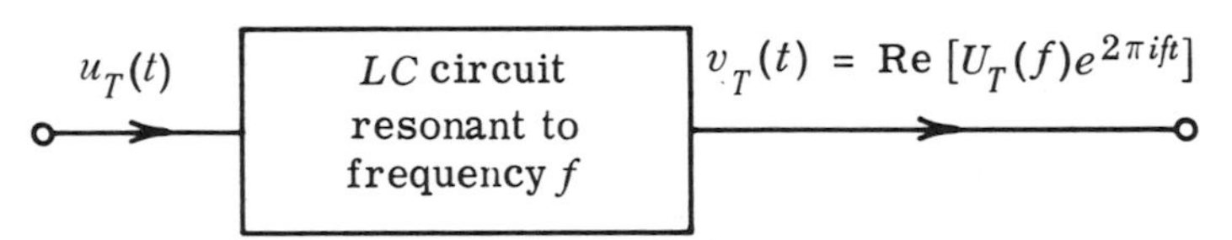


*Figure 3-2.* $U_T(f)$ is the phasor describing the response of the $LC$ circuit to $u_T(t)$ for $t \geq \frac{1}{2}T$.

as its input from $-\frac{1}{2}T$ to $\frac{1}{2}T$. Thus $U_T(f)$ indicates the amplitude and phase of the sinusoidal response of the tuned circuit, and it can therefore be regarded as a phasor describing the response of the $LC$ circuit to $u_T(t)$.

Because such a filter (or a set of such filters, tuned to different frequencies) can evidently be used in the determination of the spectrum of a random process $u(t)$, it will be useful to consider the statistics of its output amplitude $|U_T(f)|$ and, in particular, its mean-squared value, which, using (3-5), we can write as

$$\text{E}\,\{|U_T(f)|^2\} = \text{E}\left\{\int_{-\frac{1}{2}T}^{\frac{1}{2}T}\int_{-\frac{1}{2}T}^{\frac{1}{2}T} u(s)u(t)e^{2\pi if(s-t)}\,ds\,dt\right\}. \tag{3-6}$$

**The Correlation Function.** Taking the expectation before integrating, we encounter E $\{u(s)u(t)\}$, which is known as the *correlation function* (or autocorrelation function) of the random process $u(t)$. Because of the stationarity of $u(t)$, it depends only on the time difference $\tau = t - s$. Denoting it by $\psi(\tau)$, we have for any $s$ and $t$

$$\begin{aligned}\psi(\tau) &= \mathrm{E}\ \{u(s)u(s+\tau)\} \\ &= \mathrm{E}\ \{u(t-\tau)u(t)\} \\ &= \psi(-\tau);\end{aligned} \tag{3-7}$$

that is, the correlation function is always *even*. Letting $\tau = 0$, we see that the mean-squared value or average power of $u(t)$ is

$$\mathrm{E}\ \{u^2(t)\} = \psi(0). \tag{3-8}$$

We shall later evaluate the correlation functions of a number of random processes.

Now, using (3-7) in (3-6), replacing $t$ by $s + \tau$, and performing the integration with respect to $s$, we find that

$$\mathrm{E}\ \{|U_T(f)|^2\} = T \int_{-T}^{T} \left(1 - \frac{|\tau|}{T}\right) \psi(\tau) e^{-2\pi i f \tau}\, d\tau. \tag{3-9}$$

As $T \to \infty$, the first factor in the integrand approaches unity, and the integral becomes the Fourier transform of $\psi(\tau)$, which we denote by $\Psi(f)$. Thus[1]

$$\lim_{T\to\infty} \frac{E\ \{|U_T(f)|^2\}}{T} = \Psi(f). \tag{3-10}$$

## □ THE WIENER - KHINCHIN THEOREM

Returning to the case of an arbitrary linear filter with input $u(t)$ and output (3-3), we see that the average output power is

$$\mathrm{E}\ \{v^2(t)\} = \mathrm{E} \left\{ \iint_{-\infty}^{\infty} u(t-\tau_1)g(\tau_1)u(t-\tau_2)g(\tau_2)\, d\tau_1\, d\tau_2 \right\}.$$

Taking the expectation before integrating and using (3-7) with $s$ replaced by $t - \tau_2$ and $\tau = \tau_2 - \tau_1$, we find

$$\begin{aligned}\mathrm{E}\ \{v^2(t)\} &= \iint_{-\infty}^{\infty} \psi(\tau_2 - \tau_1)g(\tau_1)g(\tau_2)\, d\tau_1\, d\tau_2 \\ &= \iint_{-\infty}^{\infty} \psi(\tau)g(\tau_2 - \tau)g(\tau_2)\, d\tau_2\, d\tau,\end{aligned}$$

[1] When $\psi(\tau)$ is not absolutely integrable, we can obtain the same result by applying Parseval's theorem, (3-11), to (3-9) along with (3-25), (3-31), (3-32), Prob. 3-11, and Fig. 3-5, provided that $\psi(0) < \infty$.

which can be recognized as the integral of the product of $\psi(\tau)$ times the convolution of $g(\tau)$ with $g(-\tau)$. Now Parseval's theorem states that

$$\int_{-\infty}^{\infty} q(t)r^*(t)\,dt = \int_{-\infty}^{\infty} Q(f)R^*(f)\,df \tag{3-11}$$

for any Fourier transform pairs $q(t)$, $Q(f)$ and $r(t)$, $R(f)$, with the asterisk denoting the complex conjugate. Identifying $q(t)$ with $\psi(\tau)$ and $r(t)$ with $g(\tau) \star g(-\tau)$, we have $R(f) = G(f)G^*(f) = |G(f)|^2$ and

$$\mathrm{E}\,\{v^2(t)\} = \int_{-\infty}^{\infty} \Psi(f)|G(f)|^2\,df. \tag{3-12}$$

Being the Fourier transform of a real, even function, (3-7), $\Psi(f)$ is a real, even function too. Because $\mathrm{E}\,\{v^2(t)\}$ can never be negative and $|G(f)|^2$ is an arbitrary nonnegative even function, it follows from (3-12) that $\Psi(f)$ can never be negative—a fact confirmed by (3-10). Since $|G(f)|^2$ is the factor by which the filter attenuates power at frequency $f$, it also follows that $\Psi(f)$ must be the spectral density of the input power at frequency $f$. Because $|G(f)|^2$ is necessarily an even function, we have actually proved only that $\Psi(f) + \Psi(-f) = 2\Psi(f)$ is the power spectral density of $u(t)$ with $f$ restricted to positive values. However, it is convenient to associate half this spectral density with positive frequencies and half with negative.

Thus we have the very important *Wiener-Khinchin theorem*, which states that the Fourier transform

$$\Psi(f) = \int_{-\infty}^{\infty} \psi(\tau)e^{-2\pi i f\tau}\,d\tau \tag{3-13}$$

of the correlation function $\psi(\tau)$ of any ergodic random process is its power spectral density, and (3-12) is the average power passed by a filter having frequency response $G(f)$.

Inverting the Fourier-transform relationship (3-13), we can express the correlation function of any ergodic random process as the inverse Fourier transform of its power spectral density,

$$\psi(\tau) = \int_{-\infty}^{\infty} \Psi(f)e^{2\pi i f\tau}\,df. \tag{3-14}$$

Because the filter attenuates its power by the factor $|G(f)|^2$, the filter's output power spectral density is $\Psi_v(f) = \Psi(f)|G(f)|^2$. Taking inverse Fourier transforms, we find that the correlation function of its output is $\psi_v(\tau) = \psi(\tau) \star g(\tau) \star g(-\tau)$.

Since the statistics of a gaussian random process are completely determined by its mean and covariance function, it follows from (3-14) that the statistics of a zero-mean, ergodic, gaussian random process are likewise completely determined by its power spectrum.

From the inequalities

$$\mathrm{E}\,\{[u(t) \pm u(t+\tau)]^2\} \geq 0$$

it follows that

$$-\psi(0) \leq \psi(\tau) \leq \psi(0); \tag{3-15}$$

that is, the magnitude of the correlation function never exceeds the mean-squared value (3-8) of the random process, which is also given by (3-14) with $\tau = 0$. The correlation function of $u(t)$ is closely related to its covariance function $M(\tau) = \mathrm{E}\,\{[u(t) - \mu][u(t+\tau) - \mu]\}$, where

$$\mu = \mathrm{E}\,\{u(t)\}$$

is the mean of the random process. Applying (3-7) here, we find that

$$\psi(\tau) = M(\tau) + \mu^2. \tag{3-16}$$

**Problem 3-1.** Prove that $\psi(\tau)$ not only satisfies (3-15) but, moreover, cannot be less than $2\mu^2 - \psi(0)$.

**Other Frequency Conventions.** Because the correlation function and power spectral density are both even functions, (3-13) and (3-14) can be written

$$\Psi(f) = 2\int_0^\infty \psi(\tau)\cos 2\pi f\tau\, d\tau \tag{3-17}$$

and

$$\psi(\tau) = 2\int_0^\infty \Psi(f)\cos 2\pi f\tau\, df. \tag{3-18}$$

Some authors include in the power spectral density the factor 2 appearing in (3-18), so that the total average power is its integral over all *positive* frequencies; in this case the 2 in (3-17) becomes a 4. Some of these authors and others deal with spectral density in terms of radians per second rather than cycles per second, thereby obtaining a value $1/2\pi$ as large. Care must be exercised to ascertain these conventions before using spectral-density results. We shall use the positive-and-negative-frequency convention when dealing with the Wiener-Khinchin theorem, but later on it will be convenient to use the positive-frequency convention.

**Problem 3-2.** Prove that, if $u(t)$ and $v(t)$ are independent, jointly ergodic random processes with correlation functions and spectral densities $\psi_u(\tau)$, $\psi_v(\tau)$, $\Psi_u(f)$, and $\Psi_v(f)$, respectively, the correlation functions and spectral densities of their sum and of their product are $\psi_u(\tau) + \psi_v(\tau)$, $\psi_u(\tau)\psi_v(\tau)$, $\Psi_u(f) + \Psi_v(f)$, and $\Psi_u(f) \star \Psi_v(f)$, respectively.

**Problem 3-3.** Suppose that $u(t)$ and $v(t)$ are jointly ergodic but not necessarily independent, and let $\psi_{uv}(\tau) = \mathrm{E}\,\{u(t)v(t+\tau)\}$ and $\psi_{vu}(\tau) = \mathrm{E}\,\{v(t)u(t+\tau)\}$ denote their *crosscorrelation functions*. Show that the power spectral density of $u(t) + v(t)$ is $\Psi_u(f) + \Psi_v(f) + \Psi_{uv}(f) + \Psi_{vu}(f)$, where $\Psi_{uv}(f)$ and $\Psi_{vu}(f) = \Psi_{uv}^*(f) = \Psi_{uv}(-f)$, called the *cross-spectral densities* of $u(t)$ and $v(t)$, are the Fourier transforms of $\psi_{uv}(\tau)$ and $\psi_{vu}(\tau) = \psi_{uv}(-\tau)$, respectively, and need not be real or even functions.

**Spectral Measurement.** Often a filter like that of Fig. 3-2 is used to measure the spectrum of a random process $u(t)$ by squaring the amplitude $|U_T(f)|$ to obtain $|U_T(f)|^2$. Being a functional (3-5) of $u(t)$, however, $U_T(f)$ is a random variable and, as we shall see, despite (3-10), $|U_T(f)|^2/T$ generally does not tend toward any particular value as the measuring time $T$ grows infinite.

For example, if $u(t)$ is a zero-mean, gaussian random process, the real and imaginary parts of $U_T(f)$, Re $\{U_T(f)\}$ and Im $\{U_T(f)\}$, being linear functionals of $u(t)$ through (3-5), are jointly normal, zero-mean random variables. Moreover, $-u(t)$, $u(-t)$, and $-u(-t)$ will be just as likely to occur as $u(t)$, because of the symmetry of (1-22), and hence these two components of $U_T(f)$ will be uncorrelated and therefore statistically independent. When $T$ is large and $f \neq 0$, both components will tend to have the same variance, since $u(t)$ and $u(t - 1/4f)$ are equally likely because of the stationarity of $u(t)$, and the contributions to (3-5) coming from within $1/4f$ of $\pm\frac{1}{2}T$ are then relatively negligible; this time shift interchanges the real and imaginary components except for the end effects. (See Prob. 7-23 on p. 142.) Thus they obey the circular normal distribution (1-17), and $|U_T(f)|$ is therefore Rayleigh-distributed for large $T$. (Clearly, $U_T(0)$ is real and has a zero-mean normal distribution.)

If we increase $T$ severalfold, we get a new value for $|U_T(f)|/\sqrt{T}$ which is nearly independent of its value for the smaller (though large) $T$, because it results mainly from contributions to (3-5) that are effectively independent of those that arise with the smaller $T$. Thus, as $T$ increases, $|U_T(f)|/\sqrt{T}$ shows no tendency toward any particular value (unless toward zero for certain values of $f$), but instead takes a Rayleigh-distributed set of values with mode $\sqrt{\Psi(f)}$ in the course of its unending fluctuations. Note that the width of this distribution (Fig. 1-4) is always comparable with its central values (see Prob. 3-4$a$), i.e., it is necessarily a broad distribution. As $T$ grows, the fluctuation of $|U_T(f)|/\sqrt{T}$ slows down as $1/T$ but never diminishes in magnitude.

The same behavior of $U_T(f)$ will result whenever widely separated segments of $u(t)$ tend to be statistically independent, as in the case of the asynchronous random telegraph signal, for if $u_T(t)$ with $T$ large is regarded as a sequence of many long segments, each one will make a nearly independent contribution to $U_T(f)$, and by the central limit theorem their total will tend to be normally distributed provided that (3-8) is finite.

The failure of $|U_T(f)|/\sqrt{T}$ to approach its rms value $\sqrt{2\Psi(f)}$ for large $T$ can be observed experimentally by means of a filter with ergodic input $u(t)$ and a device for measuring the amplitude of its output, as in a spectrum analyzer. No matter how broad or sharp the frequency response of the filter, i.e., no matter how small or large the effective value of $T$, the filter's output amplitude will exhibit wide fluctuations. However, if the square of the output amplitude is averaged over a long time (e.g., by means of a square-law detector and low-pass filter), a definite, repeatable, time-average value will be obtained which, for a very sharp filter, will approximate the ensemble average (3-10) because of the ergodicity of $u(t)$.

Thus, although there is no difficulty with (3-10), $\lim_{T\to\infty} |U_T(f)|^2/T$, which has sometimes been used to define power spectral density, generally does not exist. This "limit" results from expanding $u_T(t)$ as a Fourier series whose terms are separated in frequency by $1/T$ and have coefficients of the form $U_T(f)/T$ for $f$ a multiple of $1/T$. The average power of any term, for a given sample $u(t)$, is $|U_T(f)|^2/T^2$, which gives a power spectral density $|U_T(f)|^2/T$ upon dividing by the frequency interval $1/T$.

While the ensemble average (3-7) is convenient for theoretical use, in an experimental situation only a single sample of limited duration $u_T(t)$ may be available. Here[1] it may be appropriate to approximate $\psi(\tau)$ by a time average of $u(t)u(t + \tau)$ over the available record, namely,

$$\frac{1}{T-\tau}\int_{-\frac{1}{2}T}^{\frac{1}{2}T-\tau} u(t)u(t+\tau)\,dt, \qquad \text{for } 0 \le \tau < T,$$

and to Fourier transform it to estimate $\Psi(f)$, or it may be more convenient to determine $|U_T(f)|^2/T$, which will be a rough approximation to the spectral density $\Psi(f)$. It will indeed be rough because (Prob. 3-4*a*) for large $T$ its standard deviation equals its mean.

Thus $|U_T(f)|^2/T$, as a function of $f$, will fluctuate around $\Psi(f)$ with large excursions whose widths are of the order of $1/T$, for it takes this large a change in $f$ to change (3-5) substantially. An improved approximation to $\Psi(f)$ can therefore be obtained by smoothing the curve of $|U_T(f)|^2/T$—provided that the features of $\Psi(f)$ that are of interest are broad compared to $1/T$. Alternatively, if the available record is long enough, $\Psi(f)$ can be obtained by averaging $|U_T(f)|^2/T$ with respect to $T$ or, preferably, by time-averaging with a fixed large $T$. In any case, a suitable smoothing procedure will be needed.[2] Many different weighting

[1] See R. B. Blackman and J. W. Tukey, The Measurement of Power Spectra from the Point of View of Communication Engineering, *Bell System Tech. J.*, **37**:185–282 and 485–569 (1958), reprinted by Dover Publications, Inc., New York, 1959; J. S. Bendat and A. G. Piersol, "Measurements and Analysis of Random Data," John Wiley & Sons, Inc., New York, 1966.

[2] Emanuel Parzen, On Statistical Spectral Analysis, "Stochastic Processes in Mathematical Physics and Engineering," *Proc. Symp. Appl. Math.*, **16**:221–246, American Mathematical Society, Providence, Rhode Island, 1964.

functions ("lag windows" or "spectral windows") have been proposed for this purpose, and the best one to use will depend on the particular application.

**Problem 3-4.** (*a*) From the fact that $|U_T(f)|/\sqrt{T}$ is Rayleigh-distributed for large $T$, show that $|U_T(f)|^2/T$ is exponentially distributed, i.e., that its probability density function has the form $(1/\mu)e^{-x/\mu}$ for $x > 0$ and 0 for $x \leq 0$, where $\mu$ is both its mean and its standard deviation. (*b*) For the case of the synchronous random telegraph signal, show that $|U_{T'}(f)|/\sqrt{T'}$ has a zero-mean *normal* distribution, truncated at zero, when $T'$ is large if $f$ is zero or is an odd multiple of $1/2T$, where $T$ is the length of the interval between the times of possible jumps. What is its distribution when $f$ is a multiple of $1/T$? For all other values of $f$ it is Rayleigh-distributed. (*c*) How will $|U_T(f)|/\sqrt{T}$ behave as $T \rightarrow \infty$ for a sinusoidal random process?

The results of the next two problems, along with (3-13) and (3-14), will enable us to find the output spectrum of a limiter having a gaussian input with any given spectrum.

**Problem 3-5.** Using the result of Prob. 1-7, show that, when a zero-mean stationary gaussian random process $u(t)$ with correlation function $\psi_u(\tau)$ is fed to a symmetric ideal limiter with output $v = \text{sgn } u$, the signum being 1, 0, or $-1$ accordingly as $u > 0$, $u = 0$, or $u < 0$, the correlation function of its output is

$$\psi_v(\tau) = \frac{2}{\pi} \arcsin \frac{\psi_u(\tau)}{\psi_u(0)}.$$

Show that its derivative $\psi_v'(\tau)$ has a discontinuity at $\tau = 0$ and that, consequently, $\psi_v''(0) = -\infty$.

**Problem 3-6.** Suppose that the ideal limiter in Prob. 3-5 is replaced by a symmetric error-function limiter with output $v = 2 \text{ erf } u/\sigma - 1$ varying smoothly between $-1$ and 1. By regarding $2 \text{ erf } w - 1$ as $\Pr\{x < w\} - \Pr\{x > w\} = \text{E}\{\text{sgn }(w - x)\}$, where $x$ is a standard normal random variable and "Pr" means "the probability that," and by using the result of Prob. 3-5, show that[1]

$$\psi_v(\tau) = \frac{2}{\pi} \arcsin \frac{\psi_u(\tau)}{\sigma^2 + \psi_u(0)}.$$

**The Correlation Function and Spectral Density of** $du/dt$. Given $\psi_u(\tau)$, the correlation function of $u(t)$, we can find the correlation function $\psi_{u'}(\tau)$ of its derivative $u'(t) = du(t)/dt$ by differentiating (3-7) provided,

[1] N. M. Blachman, The Correlation Function of Gaussian Noise after Error-function Limiting, *J. Electron. Control,* **16**:509–511 (1964).

of course, that this derivative exists. Thus

$$\frac{d\psi_u(\tau)}{d\tau} = \mathrm{E}\,\{u(t)u'(t+\tau)\}$$
$$= \mathrm{E}\,\{u(t-\tau)u'(t)\},$$

since $u(t)$ is stationary and hence the substitution of $t - \tau$ for $t$ does not alter this statistic. Differentiating again, we find

$$\frac{d^2\psi_u(\tau)}{d\tau^2} = -\mathrm{E}\,\{u'(t-\tau)u'(t)\}$$
$$= -\mathrm{E}\,\{u'(t)u'(t+\tau)\}$$
$$= -\psi_{u'}(\tau), \tag{3-19}$$

that is, the correlation function of $u'(t)$ is minus the second derivative of the correlation function of $u(t)$.

Substituting this result for $\psi(\tau)$ in (3-13) and integrating twice by parts, we find that the power spectral density of $u'(t)$ is

$$\Psi_{u'}(f) = 4\pi^2 f^2 \Psi_u(f). \tag{3-20}$$

This result also follows directly from (3-10) and the fact that differentiation introduces a factor $2\pi i f$ into the Fourier transform of $u_T(t)$. Equation (3-20) can be used backward to determine the spectrum of the *integral* of an ergodic random process if proper care is taken of the constant of integration.

Integration or differentiation of $u(t)$, with $\mathrm{E}\,\{u^2(t)\}$ finite, can yield a random process whose total power is infinite. In the latter case (3-19) shows that there may be a discontinuity in the slope of $\psi_u(\tau)$ at $\tau = 0$, but the derivative $u'(t)$ may nevertheless exist.

**Problem 3-7.** Show that, if $\psi_u(\tau)$ can be expanded in a power series, then

$$\psi_u(\tau) = \mathrm{E}\,\{u^2\} - \frac{\mathrm{E}\,\{u'^2\}}{2!}\tau^2 + \frac{\mathrm{E}\,\{u''^2\}}{4!}\tau^4 - \cdots.$$

**Trigonometric Series.** Regarding the trigonometric series

$$u(t) = a_0 + \Sigma a_j \cos(2\pi f_j t + \phi_j) \tag{3-21}$$

with given values for the amplitudes $a_j$, frequencies $f_j$, and phases $\phi_j$ as a sample of the ergodic process, all of those samples are translates of (3-21), that is, $u(t - t_0)$ with $t_0$ distributed uniformly between $-\frac{1}{2}T$ and $\frac{1}{2}T$ and with $T$ becoming infinite, we can calculate its correlation function as the time average

$$\psi(\tau) = \lim_{T\to\infty} \frac{1}{T} \int_{-T/2}^{T/2} u(t)u(t+\tau)\,dt. \tag{3-22}$$

Substituting (3-21) into (3-22) and integrating, we find that cross-product terms contribute nothing to the result because, as $T \to \infty$, sinusoids of different frequencies become orthogonal, and we get

$$\psi(\tau) = \lim_{T\to\infty} \frac{1}{T} \left\{ a_0^2 T + \sum a_j^2 \int_{-T/2}^{T/2} \cos(2\pi f_j t + \phi_j) \cos[2\pi f_j(t + \tau) + \phi_j]\, dt \right\}$$

$$= \lim_{T\to\infty} \left( a_0^2 + \sum \frac{a_j^2}{2T} \int_{-T/2}^{T/2} \{\cos 2\pi f_j \tau + \cos[2\pi f_j(2t + \tau) + 2\phi_j]\}\, dt \right)$$

$$= a_0^2 + \tfrac{1}{2} \sum a_j^2 \cos 2\pi f_j \tau. \tag{3-23}$$

If (3-21) is periodic, i.e., if all of the frequencies $f_j$ of the sinusoidal components of $u(t)$ are commensurable, its correlation function (3-23) is also periodic and has the same period. Any constant component $a_0$ of $u(t)$ gives rise to a constant component $a_0^2$ of $\psi(\tau)$, and a sinusoidal component of amplitude $a_j$ and frequency $f_j$, regardless of its phase, yields a cosinusoidal component of amplitude $\frac{1}{2}a_j^2$, frequency $f_j$, and phase 0. The phases of the components thus affect neither the correlation function nor the power spectrum of $u(t)$.

Fourier transforming (3-23), we get the power spectral density

$$\Psi(f) = a_0^2\delta(f) + \tfrac{1}{4}\Sigma a_j^2[\delta(f - f_j) + \delta(f + f_j)], \tag{3-24}$$

which consists entirely of discrete spectral lines at the frequencies $\pm f_j$. The delta function[1] $\delta(f - f_j)$ represents a unit impulse at $f = f_j$, that is, a function which in effect is zero except at $f = f_j$ and which encloses a unit area at that point. It is not a function in the usual sense but can be regarded as the limit of a smooth impulse of unit area and duration $T$ as $T \to 0$. However, the limit is to be taken only *after* substitution into an integral and integration. For example,

$$\int_{-\infty}^{\infty} \delta(f - f_j)g(f)\, df = g(f_j) \tag{3-25}$$

for any $g(f)$.

In general, the spectrum of a random process can be regarded as the sum of two parts, one part with a finite or continuous power spectral density and the other consisting of delta functions or spectral lines like (3-24). If the total power (3-8) of the random process is finite, the correlation function corresponding to the first part, which is its Fourier transform, will go to zero as $|\tau| \to \infty$, while that corresponding to the second part, (3-23), will not.

[1] Ron Bracewell, "The Fourier Transform and its Applications," chap. 5, McGraw-Hill Book Company, New York, 1965; Athanasios Papoulis, "The Fourier Integral and Its Applications," appendix I, p. 269, McGraw-Hill Book Company, New York, 1962.

Thus we can determine the average power of each of the sinusoidal components of any finite-power random process by examining the behavior of its correlation function for large $|\tau|$, which will always approach the form (3-23). (If the amplitudes of these components are random, $a_j^2$ will, of course, be replaced by the mean-squared amplitude of the component of frequency $f_j$, but it should be noted that in this case the random process is not ergodic.) This will be a useful technique for finding the carrier-frequency component of a frequency-modulated wave, for example, and it can also be used to determine the d-c and sinusoidal components of the output of a nonlinear device, such as a detector, which is fed a signal plus noise.

**Problem 3-8.** Using (3-22), find and plot the correlation function of a symmetric square wave of unit amplitude and frequency $F$.

## ☐ RANDOM TELEGRAPH SIGNALS

**Asynchronous Case.** For the asynchronous random telegraph signal with probability $\nu\,dt$ of a jump in any interval of length $dt$, independently of its behavior outside this interval, we can divide an interval of length $\tau$ into $\tau/dt$ increments, and we can use Bernoulli's theorem to calculate the probability of exactly $n$ jumps in the time $\tau$, namely,

$$\lim_{dt\to 0}\binom{\tau/dt}{n}(\nu\,dt)^n(1-\nu\,dt)^{(\tau/dt)-n}$$
$$=\lim_{dt\to 0}\frac{\dfrac{\tau}{dt}\left(\dfrac{\tau}{dt}-1\right)\cdots\left(\dfrac{\tau}{dt}-n+1\right)}{n!}(\nu\,dt)^n(1-\nu\,dt)^{(\tau/dt)-n}$$
$$=\frac{(\nu\tau)^n}{n!}e^{-\nu\tau}, \qquad (3\text{-}26)$$

which is the *Poisson distribution*. In (3-26) we have made use of the fact that, for any nonnegative integer $n$, both

$$\left(1-\frac{dt}{\tau}\right)\left(1-\frac{2dt}{\tau}\right)\cdots\left(1-\frac{n-1}{\tau}dt\right)$$

and $(1-\nu\,dt)^{-n}$ approach 1 as $dt\to 0$, while $(1-\nu\,dt)^{-1/(\nu dt)}$ becomes $e$. The symbol $\binom{m}{n}$ stands for the binomial coefficient

$$\frac{m!}{n!(m-n)!}=m(m-1)\cdots\frac{m-n+1}{n!}.$$

Note that the summation of (3-26) over all nonnegative integers $n$ is unity.

Since the product $u(t)u(t+\tau)$ is 1 or $-1$ accordingly as the number of jumps in the interval from $t$ to $t+\tau$ is even or odd, we can use (3-26) to find the correlation function of the asynchronous random telegraph signal $u(t)$ by subtracting the total probability of an odd number of jumps during a time $|\tau|$ from that of an even number of jumps. However, it is interesting to use a less direct approach that avoids the need of any calculation.

We suppose that the sequence of jumps is generated by a Poisson process having the rate $2\nu$ and that after each event of that process a

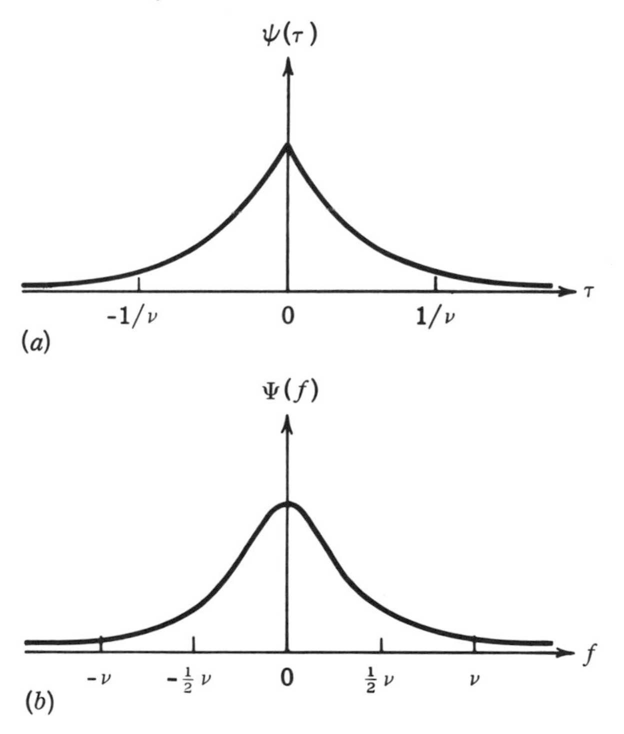


**Figure 3-3.** (*a*) Correlation function and (*b*) power spectral density of the asynchronous random telegraph signal.

fair coin is tossed to determine whether or not $u(t)$ is to have a jump at that time. The resulting jumps clearly form a Poisson process of rate $\nu$, and evidently, if the coin is tossed between times $t$ and $t + \tau$, then $\mathrm{E}\,\{u(t)u(t+\tau)\} = 0$, while, if it is not, $\mathrm{E}\,\{u(t)u(t+\tau)\} = 1$. The probability of the latter event, given by (3-26) with $n = 0$ and $\nu$ replaced by $2\nu$, is thus the correlation function

$$\psi(\tau) = e^{-2\nu|\tau|}, \tag{3-27}$$

which is shown in Fig. 3-3*a*. Figure 3-3*b* shows its Fourier transform, the power spectral density

$$\Psi(f) = \frac{\nu}{\nu^2 + \pi^2 f^2}. \tag{3-28}$$

This third-order curve is known as the "witch of Agnesi," because it was named "versiera" by the Italian mathematician Maria Gaetana Agnesi (1718–99), to whom it looked like a witch's hat.

**Problem 3-9.** Find the correlation function of the asynchronous random telegraph signal by the direct method, and show that (3-28) is its Fourier transform.

**Problem 3-10.** Show that the mean and variance of the Poisson distribution (3-26) are both equal to $\nu\tau$.

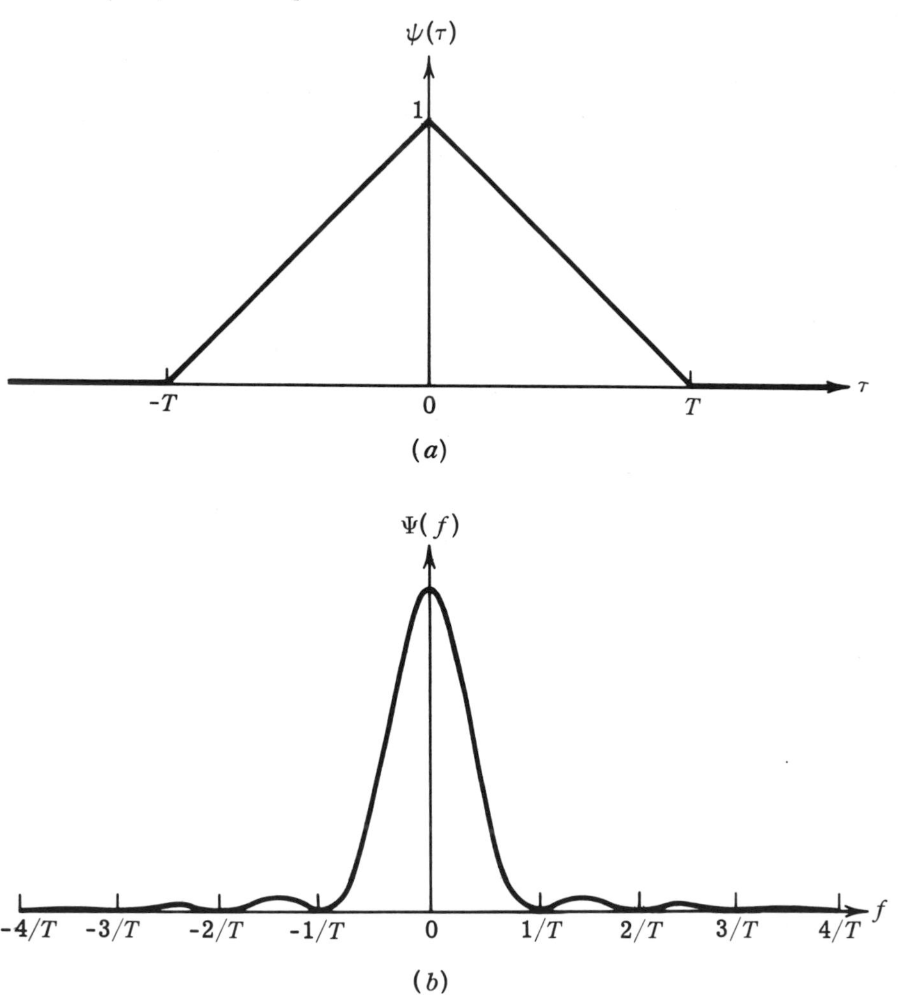


*Figure 3-4.* (*a*) Correlation function and (*b*) power spectral density of the synchronous random telegraph signal.

**Synchronous Case.** For the synchronous random telegraph signal $u(t)$ we likewise have $E\{u(t)u(t+\tau)\} = 0$ if the end of a clock period falls between times $t$ and $t + \tau$, and $E\{u(t)u(t+\tau)\} = 1$ if the times $t$ and $t + \tau$ fall within the same clock interval. If $|\tau|$ exceeds the clock-period duration $T$, the former will certainly be the case, and $\psi(\tau) = 0$. For $|\tau| \leq T$ the probability that the end of a clock period falls between

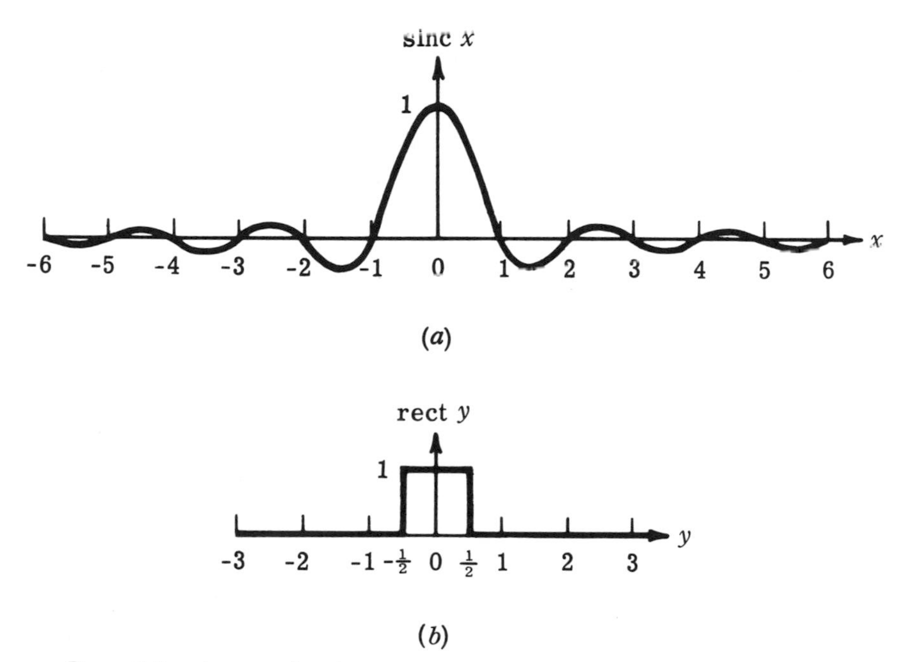


*Figure 3-5.* sinc $x$ and rect $y$.

$t$ and $t + \tau$ is $|\tau|/T$ because of the uniform distribution of the clock epoch. Hence the correlation function is

$$\psi(\tau) = \begin{cases} 1 - \dfrac{|\tau|}{T} & \text{for } |\tau| \leq T, \\ 0 & \text{for } |\tau| \geq T, \end{cases} \tag{3-29}$$

which is shown in Fig. 3-4*a*. Its Fourier transform, shown in Fig. 3-4*b*, is the power spectral density

$$\Psi(f) = T \operatorname{sinc}^2 fT, \tag{3-30}$$

where[1] (Fig. 3-5)

$$\operatorname{sinc} x = \frac{\sin \pi x}{\pi x} \tag{3-31}$$

[1] P. M. Woodward, "Probability and Information Theory with Applications to Radar," chap. 2, p. 26, Pergamon Press, New York, 1953; Bracewell, *op. cit.*, p. 62.

is the Fourier transform of

$$\operatorname{rect} y = \begin{cases} 1 & \text{for } |y| < \frac{1}{2}, \\ 0 & \text{for } |y| > \frac{1}{2}. \end{cases} \tag{3-32}$$

**Problem 3-11.** Obtain (3-30) from (3-29) and the Wiener-Khinchin theorem by regarding (3-29) as $1/T$ times the convolution of rect $\tau/T$ with itself.

**Problem 3-12.** Generalizing the synchronous random telegraph signal, put $u(t) = \Sigma a_j v(t - t_j)$, with $t_j = jT + \theta$, $\theta$ uniformly distributed between 0 and $T$, the $a_j$ independent random variables having mean zero and variance $\sigma^2$, and $v(t)$ being an arbitrary finite-energy waveform having Fourier transform $V(f)$. Show that the power spectral density of $u(t)$ is $\Psi(f) = \sigma^2|V(f)|^2/T$, and apply this result to the synchronous random telegraph signal.

**Problem 3-13.** Show that the power spectral density of $u(t) = \Sigma a_j v(t - t_j)$, with $t_j$ the times of the events of a Poisson process of rate $\nu$, the $a_j$ independent random variables having mean $\mu$ and variance $\sigma^2$, and $v(t)$ as in Prob. 3-12, is $\Psi(f) = (\sigma^2 + \mu^2)\nu|V(f)|^2 + \mu^2\nu^2V^2(0)\delta(f)$. Note that the asynchronous random telegraph signal is *not* a special case of this $u(t)$ and that for $\mu = 0$, as $\nu \to \infty$ and $\sigma \to 0$, by the central limit theorem $u(t)$ becomes an ergodic gaussian random process.[1]

## □ AMPLITUDE MODULATION

In Prob. 3-2 we saw that the correlation function of the product of two jointly ergodic random processes is the product of their correlation functions. An important special case of this amplitude modulation of one random process by another is that in which one factor is a sinusoidal random process of fixed amplitude, that is, $\cos(2\pi Ft + \theta)$, with $\theta$ uniformly distributed between 0 and $2\pi$. Thus the correlation function of the

[1] Such a representation of gaussian noise can be useful in calculating the effect of this noise on a nonlinear or varying system, since it may be easy to determine the effect of an individual small pulse $a_j v(t - t_j)$, and from it, the statistical effect of the entire random train of pulses. When the noise is "white," i.e., has a power spectral density that is constant over all frequencies of interest, we can take $v(t) = \delta(t)$. Although the pulses then cannot overlap, and their sum cannot be gaussian, the system's response to them will be the same as if they had passed through a filter that turned them into overlapping pulses and their sum into gaussian noise just as if $u(t)$ had been gaussian noise. See R. L. Stratonovich, "Topics in the Theory of Random Noise," vol. 1, chap. 4 (translated by R. A. Silverman), Gordon and Breach, Science Publishers, Inc., New York, 1963; and N. M. Blachman, The Effect of Noise on a Nonlinear Control System, in J. F. Coales, *et al.* (eds.), "Automatic and Remote Control," vol. 2, pp. 770–773, Butterworth & Co. (Publishers), Ltd., London, 1961.

amplitude-modulated sinusoid

$$v(t) = u(t) \cos (2\pi Ft + \theta) \tag{3-33}$$

is (see Prob. 3-14 below)

$$\psi_v(\tau) = \tfrac{1}{2}\psi_u(\tau) \cos 2\pi F\tau, \tag{3-34}$$

provided that the ergodic random process $u(t)$, with correlation function $\psi_u(\tau)$, contains no periodicity of any sort commensurable with $1/F$, the reciprocal of the carrier frequency. The power spectral density of (3-33) is the Fourier transform of (3-34),

$$\Psi_v(f) = \tfrac{1}{4}\Psi_n(f - F) + \tfrac{1}{4}\Psi_u(f + F). \tag{3-35}$$

Thus, for a sufficiently large carrier frequency, the spectrum of the modulated wave consists of two well separated parts corresponding to the two terms of (3-35), the first symmetric about $f = F$ and the second about $f = -F$, and both parts have the same shape as the spectrum of the modulating waveform.

**Problem 3-14.** Show that the sinusoidal carrier $\cos (2\pi Ft + \theta)$ with $\theta$ uniformly distributed between 0 and $2\pi$ is ergodic and has correlation function $\frac{1}{2} \cos 2\pi F\tau$. What is the significance of the fact that this is the same as the correlation function (2-1) of the sinusoidal gaussian random process with $\sigma^2 = \frac{1}{2}$?

**Random Telegraphic Modulation.** If, for example, the modulating wave $u(t)$ is an asynchronous random telegraph signal, the modulated wave $v(t)$ is a sinusoid $\pm \cos (2Ft + \theta)$ whose phase is changed by 180° at the instants given by a Poisson process, and its spectrum is the sum of two witches of Agnesi. Similarly, if $u(t)$ is a synchronous random telegraph signal, $v(t)$ has 180° changes of phase with probability $\frac{1}{2}$ at regularly spaced times, and its spectrum consists of two parts having the shape shown in Fig. 3-4*b*, one shifted $F$ to the right and the other $F$ to the left. However, if $FT$ is rational, there is a fixed phase relationship in any one sample $v(t)$ between the modulation and the carrier, and $v(t)$ is not ergodic. In this case the foregoing spectrum is only the average over all possible phase relationships, for the time statistics of different samples will yield different spectra.

**Problem 3-15.** Let $u(t)$ be a synchronous random telegraph signal whose sign changes with probability $\frac{1}{2}$ at the times $\ldots, -T + \theta, \theta, T + \theta, 2T + \theta, \ldots$, where $\theta$ is a random variable distributed uniformly between 0 and $T$. Find the spectrum of $v(t) = u(t) \cos [2\pi F(t - \theta) + \phi]$, with $\phi$ fixed and $FT = k$ an integer. Explain why the spectrum of this ergodic random process differs from the result of using (3-30) in (3-35), and reconcile the two.

## ☐ ANGLE - MODULATED WAVES

The "instantaneous frequency" of a wave of the form $A \cos \phi(t)$ may be defined as $(2\pi)^{-1}\, d\phi/dt$. Thus, if a random process $u(t)$ frequency modulates a sinusoidal carrier of amplitude $A$ and frequency $F$ so that the instantaneous frequency of the modulated wave $v(t)$ is $F + u(t)$, we can write

$$v(t) = A \cos \left[2\pi F t + 2\pi \int_0^t u(s)\, ds + \theta\right] \tag{3-36}$$

for some $\theta$. (Any other fixed lower limit for the integral would do just as well as zero; changing it is equivalent to changing $\theta$.) If $u(t)$ is ergodic and involves no periodicity commensurable with $1/F$, and if $\theta$ is distributed uniformly between 0 and $2\pi$, that is, if the modulation and the carrier are jointly ergodic, then $v(t)$ is ergodic too, although the integral of $u(t)$ generally is not.

**Correlation Function and Spectrum.** The correlation function $\psi_v(\tau) = \mathrm{E}\,\{v(t)v(t+\tau)\}$ is the expectation of the product of two cosines, which can be expressed as half the cosine of the difference of their arguments plus half the cosine of the sum. Since the latter includes $2\theta$, the expectation of the second term is zero, and we are left with

$$\psi_v(\tau) = \tfrac{1}{2}A^2\, \mathrm{E}\left\{\cos\left[2\pi F\tau + 2\pi \int_t^{t+\tau} u(s)\, ds\right]\right\}. \tag{3-37}$$

Denoting this integral, which for fixed $\tau$ is an ergodic random process (Prob. 2-10), by

$$w = \int_t^{t+\tau} u(s)\, ds, \tag{3-38}$$

and expressing the cosine as half the sum of two imaginary exponentials, we can write (3-37) in the form

$$\begin{aligned} \psi_v(\tau) &= \tfrac{1}{4}A^2\, \mathrm{E}\,\{e^{2\pi iF\tau + 2\pi iw} + e^{-2\pi iF\tau - 2\pi iw}\} \\ &= \tfrac{1}{4}A^2 z(\tau) e^{2\pi iF\tau} + \tfrac{1}{4}A^2 z^*(\tau) e^{-2\pi iF\tau}, \end{aligned} \tag{3-39}$$

where

$$z(\tau) = \mathrm{E}\,\{e^{2\pi iw}\}. \tag{3-40}$$

Although $w$ depends on $t$, $\tau$, and the particular form taken by $u(t)$, because of the stationarity of $w$ as a function of $t$ (3-40) depends only on $\tau$.

Expressing $z(-\tau)$ by means of (3-40) and replacing $t$ by $t + \tau$, we see that $z(-\tau) = z^*(\tau)$. Hence the Fourier transform $Z(f)$ of $z(\tau)$ is real. Fourier transforming (3-39), we get the power spectral density

$$\Psi_v(f) = \tfrac{1}{4}A^2 Z(f - F) + \tfrac{1}{4}A^2 Z(-f - F). \tag{3-41}$$

Thus, for large $F$, the spectrum of $v(t)$ consists of two peaks, one near $f = F$ and the other its mirror image in the line $f = 0$. In contrast to the amplitude-modulation case, where each peak is symmetric about the carrier frequency, the peaks of (3-41) need not be at all symmetric.

**Problem 3-16.** Show that[1] if $u(t) = \Sigma u_j(t)$ is the sum of statistically independent, jointly ergodic (along with the carrier) random processes $u_j(t)$, to each of which corresponds a $z_j(\tau)$ as in (3-40), then $Z(f)$ is the convolution of their Fourier transforms. In particular, if one component of $u(t)$ is periodic, the spectrum of $v(t)$ is the sum of the spectra resulting from frequency modulation by the remainder of $u(t)$ of each of the Fourier components of the carrier when modulated by the periodic component alone.

**Spectral Width.** In general, the spectrum (3-41) is difficult to evaluate, but its "rms width"

$$\rho = \sqrt{\frac{\int_{-\infty}^{\infty} f^2 \Psi_v(f)\, df}{\int_{-\infty}^{\infty} \Psi_v(f)\, df}}, \tag{3-42}$$

which is the radius of gyration of the area under $\Psi_v(f)$, is quite easy to determine. From (3-8) and (3-14) we see that the denominator of (3-42) is $\mathrm{E}\,\{v^2(t)\} = \psi_v(0) = \frac{1}{2}A^2$, by (3-37).

From (3-20) we see that the integrand in the numerator of (3-42) is $1/4\pi^2$ times the power spectral density of

$$v'(t) = -2\pi A[F + u(t)] \sin \left[2\pi F t + 2\pi \int_0^t u(s)\, ds + \theta\right]. \tag{3-43}$$

By (3-10), then, the numerator is $(4\pi^2)^{-1}\,\mathrm{E}\,\{v'^2(t)\}$, which, on averaging the square of (3-43) with respect to $\theta$, is seen to be $\frac{1}{2}A^2\,\mathrm{E}\,\{[F + u(t)]^2\}$. Substituting into (3-42), we find

$$\rho = \sqrt{\mathrm{E}\,\{[F + u(t)]^2\}}, \tag{3-44}$$

that is, the radius of gyration of the spectrum of a frequency-modulated sinusoid is the rms instantaneous frequency.

When $F$ is large and the two spectral peaks described by the two terms of (3-41) are well separated, it is of more interest to know the width of each peak separately than of the spectrum as a whole. For this purpose we substitute (3-41) into the numerator of (3-42), replace the denominator by $\frac{1}{2}A^2$, and substitute (3-44) for $\rho$, thus finding that

$$\mathrm{E}\,\{[F + u(t)]^2\} = \tfrac{1}{2} \int_{-\infty}^{\infty} f^2[Z(f - F) + Z(-f - F)]\, df.$$

Expanding the left-hand side, noticing that the integrals of both terms on the right-hand side are equal, and replacing $f$ in the first term by

[1] J. W. Goodman, Power Spectra of Waves with Multiple Angle Modulations, *Proc. IEEE*, **52**:418 (1964).

$f + F$, we get

$$F^2 + 2F \, \mathrm{E} \, \{u(t)\} + \mathrm{E} \, \{u^2(t)\} = \int_{-\infty}^{\infty} (F + f)^2 Z(f) \, df.$$

Since this equation holds for all values of $F$, the coefficients of the various powers of $F$ on its two sides must be respectively equal. Thus

$$\int_{-\infty}^{\infty} Z(f) \, df = 1, \tag{3-45}$$

$$\int_{-\infty}^{\infty} f Z(f) \, df = \mathrm{E} \, \{u(t)\}, \tag{3-46}$$

$$\int_{-\infty}^{\infty} f^2 Z(f) \, df = \mathrm{E} \, \{u^2(t)\}. \tag{3-47}$$

Dividing (3-46) by (3-45), we see that the centroid of $Z(f)$ is given by the average frequency deviation $\mathrm{E} \, \{u(t)\}$. Dividing (3-47) by (3-45), we find that the second moment of $Z(f)$ equals the mean squared frequency deviation $\mathrm{E} \, \{u^2(t)\}$. Hence the radius of gyration or "rms width" of each of the two parts of the spectrum of a frequency-modulated sinusoid is the rms difference between the instantaneous frequency and its average value, namely,

$$\begin{aligned} \rho &= \sqrt{\frac{\int_{-\infty}^{\infty} (f - \mu)^2 Z(f) \, df}{\int_{-\infty}^{\infty} Z(f) \, df}} \\ &= \sqrt{\mathrm{E} \, \{[u(t) - \mu]^2\}}, \end{aligned} \tag{3-48}$$

which is not to be confused with (3-42) and (3-44), for which we used the same symbol $\rho$; here

$$\begin{aligned} \mu &= \frac{\int_{-\infty}^{\infty} f Z(f) \, df}{\int_{-\infty}^{\infty} Z(f) \, df} \\ &= \mathrm{E} \, \{u(t)\}. \end{aligned} \tag{3-49}$$

A waveform of rms bandwidth $\rho$, (3-48), necessarily occupies a band of width at least $2\rho$, for otherwise the radius of gyration of its spectrum would be less than $\rho$. On the other hand, at least three-quarters of its total power must fall within a band of width $4\rho$, for otherwise the radius of gyration of its spectrum would exceed $\rho$. In the case of a rectangular spectrum (with spectral density constant over a band of frequencies and zero outside it), the bandwidth is $2\sqrt{3}\,\rho$, and in general a decision as to the bandwidth required for reasonably undistorted transmission must be based on the exact shape of the spectrum, since $\rho$ gives only a rough idea of it.

**Problem 3-17.** Prove (3-45), (3-46), and (3-47) by expressing their left-hand sides in terms of (3-40) and its first two derivatives for $\tau = 0$.

**Gaussian Modulation.** When $u(t)$ is a zero-mean, ergodic, gaussian random process, $w = \int_t^{t+\tau} u(s)\, ds$ for fixed $t$, and $\tau$, being a sum of values of $u(t)$, is a zero-mean gaussian random variable. To evaluate $z(\tau)$, (3-40), we shall need the variance of $w$, which, with the help of (3-32), can be expressed as

$$w = \int_{-\infty}^{\infty} u(s) \operatorname{rect} \frac{t + \frac{1}{2}\tau - s}{\tau}\, ds,$$

that is, as the convolution of $u(t)$ with rect $(t + \frac{1}{2}\tau)/\tau$. Hence $w$ is the result of passing $u(t)$ through a linear filter with impulse response $g(t) = \operatorname{rect}\,(t + \frac{1}{2}\tau)/\tau$. Its Fourier transform, $G(f) = e^{\pi i f \tau}|\tau| \operatorname{sinc} f\tau$, is the frequency response of the filter. Thus, from (3-12), the variance of the ergodic random process $w$ is

$$\mathrm{E}\,\{w^2\} = \tau^2 \int_{-\infty}^{\infty} \Psi_u(f) \operatorname{sinc}^2 f\tau\, df. \tag{3-50}$$

Recognizing (3-40) as the characteristic function of the univariate distribution of $w$ with argument $2\pi$, we obtain $z(\tau)$ from (1-27) by setting $z = 2\pi$, $\mu = 0$, and $\sigma^2$ equal to (3-50), namely,

$$\begin{aligned} z(\tau) &= \exp\,(-2\pi^2\, \mathrm{E}\,\{w^2\}) \\ &= \exp\left[-2\pi^2\tau^2 \int_{-\infty}^{\infty} \Psi_u(f) \operatorname{sinc}^2 f\tau\, df\right]. \end{aligned} \tag{3-51}$$

Since this $z(\tau)$ is an even function, its Fourier transform $Z(f)$ is also even, and the two parts of the spectrum (3-41) are symmetric about $f = \pm F$. Substituting (3-51) into (3-39), we get the correlation function of (3-36) for the case of zero-mean gaussian frequency modulation,

$$\psi_v(\tau) = \tfrac{1}{2}A^2 \cos 2\pi F\tau \exp\left[-2\int_{-\infty}^{\infty} \Psi_u(f) \sin^2 \pi f\tau \frac{df}{f^2}\right]. \tag{3-52}$$

The spectral density of $v(t)$, the Fourier transform of (3-52), can almost never be found analytically (but see Prob. 3-18 below), and approximations are needed in its evaluation. However, we shall find it easy to determine from (3-52) the strength of the carrier-frequency component of the spectrum of $v(t)$.

**Problem 3-18.** Show that the power spectral density of a sinusoid of amplitude $A$ and frequency $F$ frequency-modulated by a gaussian random process with spectral density $\Psi_u(f) = a|f|e^{-b|f|}$ is given by (3-41) with $Z(f) = [b/\sqrt{\pi}\,\Gamma(a)](\frac{1}{2}b|f|)^{a-\frac{1}{2}}K_{a-\frac{1}{2}}(b|f|)$, where $K_{a-\frac{1}{2}}$ is the modified Bessel function of the third kind. Notice that, in the neighborhood of $f = 0$, $Z(f)$ is of the order of $|f|^{2a-1}$ for $a < \frac{1}{2}$, $\ln |f|$ for $a = \frac{1}{2}$, and unity for $a > \frac{1}{2}$, and that, for $a = 1$, $Z(f) = \frac{1}{2}be^{-b|f|}$. (HINT: See Arthur Erdélyi and the Bateman Project Staff, "Tables of Integral Transforms," vol. 1, p. 157, No. (17), and p. 11, No. (7), McGraw-Hill Book Company, New York, 1954.)

**Carrier-Frequency Spectral Line.** By reference to (3-23) we see that any sinusoidal component of $v(t)$ will manifest itself in the correlation function (3-52) as a cosinusoidal component of the same frequency, which does not vanish as $|\tau| \to \infty$. Thus we can determine the magnitude of the carrier-frequency component of $v(t)$ in the case of zero-mean gaussian frequency modulation by examining the behavior of (3-52) in this limit.

The larger $|\tau|$ is, the more rapidly the sine in (3-52) oscillates, and in the limit the integral approaches the value obtained by replacing the square of the sine by its average value, $\frac{1}{2}$. For sufficiently large $|\tau|$, then, (3-52) is approximately

$$\psi_v(\tau) = \tfrac{1}{2}A^2 \cos 2\pi F\tau \exp\left[-\int_{-\infty}^{\infty} \Psi_u(f)\, \frac{df}{f^2}\right] \tag{3-53}$$

if this integral is finite. If the integral is infinite, as it always is when $\Psi_u(0) \neq 0$, the exponential of (3-52) falls to zero as $|\tau| \to \infty$, and there is no carrier-frequency component of $v(t)$, for the low-frequency components of the modulating wave then cause the phase of the cosine in (3-36) to wander indefinitely far from $\theta$, destroying any long-term phase coherence. In fact, the mean-squared departure of the phase from $\theta$ equals this integral (see Prob. 3-19 below).

By comparing (3-53) with (3-23), we see that, when this integral is finite, $v(t)$ contains a discrete spectral line with power $\frac{1}{4}A^2 \exp\left[-\int_{-\infty}^{\infty} \Psi_u(f)\, df/f^2\right]$ on each of the frequencies $\pm F$. This exponential is thus the factor by which the modulation attenuates the carrier-frequency component.

**Problem 3-19.** Show that the correlation function of a sinusoid $v(t) = A \cos\,[2\pi Ft + u(t) + \theta]$ that is *phase-modulated* by an ergodic gaussian $u(t)$ with correlation function $\psi_u(\tau)$ and power spectral density $\Psi_u(f)$ is $\psi_v(\tau) = \frac{1}{2}A^2 \cos 2\pi F\tau \exp\,[\psi_u(\tau) - \psi_u(0)] = \frac{1}{2}A^2 \cos 2\pi F\tau \exp\left[-2\int_{-\infty}^{\infty} \Psi_u(f) \sin^2 \pi f\tau\, df\right]$ and compare with (3-52). Find and interpret $\psi_v(\tau)$ for large $|\tau|$. If $\exp \psi_u(\tau)$ is expanded as a power series in $\psi_u(\tau)$, then, on taking the Fourier transform of $\psi_v(\tau)$, we obtain for $\Psi_v(f)$ a convergent series of positive terms which are convolutions of the spectrum of $u(t)$ with itself.[1]

[1] N. M. Abramson, Bandwidth and Spectra of Phase- and Frequency-Modulated Waves, *IEEE Trans. Comm. Systems*, **CS-11**:407–414 (December, 1963).

**The Effect of Weak Modulation.** The integral in (3-53) is finite if E $\{u^2(t)\} < \infty$ and if $\Psi_u(f)$ is of a smaller order than $f$ in the neighborhood of $f = 0$. Thus it is the lowest-frequency components of the modulating wave $u(t)$ that account for the disappearance of the discrete carrier-frequency component of $v(t)$ when $\Psi_u(0)$ is not zero. When $\Psi_u(0)$ is small, the carrier-frequency spectral line is spread into a narrow, sharply peaked continuous spectrum, whose shape we can readily determine.

With $\Psi_u(f)$ small, (3-51) is approximately unity unless $|\tau|$ is very large. In the latter case, $|\tau| \operatorname{sinc}^2 f\tau$ (Fig. 3-4*b*) becomes a narrow unit impulse equivalent to $\delta(f)$, which [see (3-25)] gives us

$$z(\tau) = \exp\left[-2\pi^2\Psi_u(0)|\tau|\right]. \tag{3-54}$$

Comparing (3-54), which is evidently accurate for small $|\tau|$ as well as for large, with (3-27), we see, with the help of (3-34) and (3-39), that the spectrum of $v(t)$ is the same as that of a sinusoid of amplitude $A$ and frequency $F$ amplitude-modulated (i.e., multiplied) by an asynchronous random telegraph signal with $\nu = \pi^2\Psi_u(0)$, namely,

$$\Psi_v(f) = \frac{\frac{1}{4}A^2\Psi_u(0)}{\pi^2\Psi_u{}^2(0) + (f - F)^2} + \frac{\frac{1}{4}A^2\Psi_u(0)}{\pi^2\Psi_u{}^2(0) + (f + F)^2}. \tag{3-55}$$

Since the derivation of (3-55) involves only the statistics of $w$, (3-38), for large $|\tau|$, with $w$ the sum of many values of $u(t)$ over a long interval, this same witch-of-Agnesi spreading of the carrier-frequency spectral line will, by the central limit theorem, occur with many kinds of random frequency modulation—nongaussian as well as gaussian.

From (3-48) it is evident that, when the modulation $u(t)$ has a finite average power, (3-55) cannot hold for all values of $f$, since the witch of Agnesi has an infinite rms width. In fact, our derivation holds only over a range of frequencies near $f = \pm F$ roughly equal in width to the range of frequencies near $f = 0$ over which $\Psi_u(f)$ is approximately $\Psi_u(0)$. Thus, as the modulation is attenuated and the witches become narrower, (3-55) becomes accurate farther down their sides.

Having found the effect of modulation whose spectral density is small at low frequencies, we can use the result of Prob. 3-16 to find the effect of higher-frequency modulation components, which need not be weak. They determine the tails of the spectrum of $v(t)$, which may be more important than the high but narrow witches of Agnesi (3-55).

The case of $\Psi_u(0) = 0$, to which (3-55) does not apply, is generally better treated in terms of *phase* modulation (Prob. 3-19), for phase modulation by $u(t)$ with spectral density $\Psi_u(f)$ is equivalent to frequency modulation by $(2\pi)^{-1}\, du(t)/dt$ with spectral density $f^2\Psi_u(f)$.

**Problem 3-20.** Show that phase modulation by any ergodic random process $u(t)$ whose values are nearly always small yields the same spectrum as amplitude modulation that multiplies the carrier by $1 + u(t)$. Note that, since $A \cos [2\pi Ft + u(t) + \theta]$ is approximately $A \cos (2\pi Ft + \theta) - Au(t) \sin (2\pi Ft + \theta)$, the only difference is a 90° phase change in the sidebands.

**Slow, Strong Frequency Modulation.** When the frequency swing of the carrier is large compared to the frequencies present in the modulation, i.e., when the modulation is slow and strong and the modulation index is large, it is easy to approximate the power spectral density (3-41) of (3-36), regardless of the statistics of the modulation $u(t)$. Since $u(t)$ varies slowly, if $|\tau|$ is not too large we have from (3-38) the approximation $w = u(t)\tau$.

Substituting this value into (3-40) and determining the average by multiplying by the univariate probability density function $p(u)$ of $u(t)$ and integrating, we have

$$z(\tau) = \int_{-\infty}^{\infty} p(u)e^{2\pi i\tau u}\,du, \tag{3-56}$$

which is the inverse Fourier transform of $p(u)$. Hence the Fourier transform of $z(\tau)$ is $Z(f) = p(f)$, and (3-41) is

$$\Psi_v(f) = \tfrac{1}{4}A^2p(f - F) + \tfrac{1}{4}A^2p(-f - F). \tag{3-57}$$

The approximation (3-56) is good only when $|\tau|$ is small compared to the reciprocal of the highest significant frequency $b$ in the spectrum of the modulating wave $u(t)$, but when the modulation is sufficiently strong and has a continuous univariate distribution, even the smallest features of its probability density function have a breadth $B$ that is large compared to $b$. Hence (3-56), which is accurate for $|\tau| \ll 1/b$, is very small for $1/B \ll |\tau| \ll 1/b$. Although (3-56) remains small for larger $|\tau|$, in some cases $z(\tau)$ does not remain small; e.g., with periodic modulation $z(\tau)$ will be large for $\tau$ near any multiple of its period. In such cases (3-57) lacks the fine structure (spectral lines in the case of periodic modulation) contributed by such values of $z(\tau)$ and is, in effect, smoothed over bands of frequencies of the order of the spectral width $b$ of the modulation.

Alternatively, (3-57) can be derived from the fact that $\frac{1}{2}A^2p(f - F)\,df$ would be the power passed by a filter of bandwidth $df$ at frequency $f$, since the slowness of the modulation ensures that the filter passes $v(t)$ during the fraction of the time that $F + u(t)$, the instantaneous frequency, lies within its pass band.

**The Effect of AM Along with Strong FM.** Equation (3-57), variously called the *adiabatic theorem* and *Woodward's theorem*,[1] asserts that

[1] P. M. Woodward, "The Spectrum of Random Frequency Modulation," Telecommunications Research Establishment, Great Malvern, Worcestershire, Memo-

for slow FM the spectrum is given by the univariate distribution of instantaneous frequencies. If the carrier is modulated slowly in amplitude $A$ as well as in frequency, its spectrum will be approximately just the average of (3-57) over the distribution of $A$, which need not be statistically independent of the frequency deviation $u$. Thus, denoting the conditional distribution of $u$ for a given $A$ by $p(u|A)$ and the probability density function of $A$ by $p(A)$, we obtain[1]

$$\begin{aligned}\Psi_v(f) &= \tfrac{1}{4}\int_0^\infty A^2 p(A)[p(f - F|A) + p(-f - F|A)]\,dA \\ &= \tfrac{1}{4}\int_0^\infty A^2[p(f - F,A) + p(-f - F,A)]\,dA,\end{aligned} \tag{3-58}$$

where $p(u,A) = p(A)p(u|A)$ is the joint probability density function of the instantaneous frequency deviation and amplitude.[2]

## ☐ THE ZEROS OF A GAUSSIAN RANDOM PROCESS

Except in certain special cases, the problem of determining the probability distribution of the interval between successive zeros of an ergodic gaussian random process $u(t)$ with correlation function $\psi(\tau)$ and power spectral density $\Psi(f)$ remains unsolved. However, it is easy to find the expected number of zeros per unit time by making use of the bivariate normal distribution of $u(t)$ and its derivative $u'(t)$ for a fixed $t$. (See Fig. 3-6.)

The mean value of $u'$ is the derivative of the (constant) mean value of $u$, namely, zero, and we suppose that the mean value of $u$, too, is zero. These two random variables are uncorrelated, for

$$\mathrm{E}\,\{uu'\} = \tfrac{1}{2}(d/dt)\,\mathrm{E}\,\{u^2\},$$

being the derivative of a constant, is zero. Denoting this constant by $\sigma^2 = \mathrm{E}\,\{u^2\} = \psi(0) = \int_{-\infty}^{\infty} \Psi(f)\,df$, with the help of (3-19) and (3-20) we can express the variance of $u'$ as

$$\mathrm{E}\,\{u'^2\} = -\psi''(0) = 4\pi^2 \int_{-\infty}^{\infty} f^2\Psi(f)\,df = 4\pi^2\rho^2\sigma^2,$$

where $\rho$ is the "rms width" or radius of gyration of the spectrum of $u(t)$, defined as in (3-42). Substituting these moments into (1-12), we obtain

---

randum No. 666, December, 1952; N. M. Blachman, Limiting Frequency-Modulation Spectra, *Information and Control*, **1**:26–37 (1957); J. A. Mullen and D. Middleton, Limiting Forms of FM Noise Spectra, *Proc. IRE*, **45**:874–877 (1957).

[1] N. M. Blachman, A Generalization of Woodward's Theorem on FM Spectra, *Information and Control*, **5**:55–63 (1962).

[2] Further results for simultaneous amplitude and phase modulation can be found in D. Middleton, "An Introduction to Statistical Communication Theory," sec. 14.3, McGraw-Hill Book Company, New York, 1960.

the joint probability density function of $u$ and $u'$,

$$\frac{1}{4\pi^2\rho\sigma^2}\exp\left(-\frac{u^2}{2\sigma^2}-\frac{u'^2}{8\pi^2\rho^2\sigma^2}\right). \tag{3-59}$$

We shall now use (3-59) to find the probability that $u(t)$ has an upward zero crossing in the infinitesimal time interval between $t$ and $t + dt$. For this to occur, $u(t)$ must be negative, and $u'(t)$ must be large enough that $u(t) + u'(t)\,dt \cong u(t + dt)$ is positive; that is, we must have $-u'\,dt < u < 0$. To find the probability that both of these inequalities are satisfied, we first integrate (3-59) with respect to $u$ from $-u'\,dt$ to

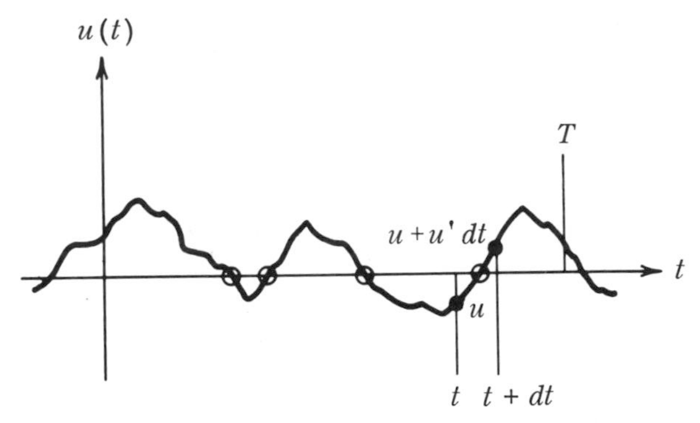


*Figure 3-6.* The zeros of $u(t)$ in the interval $(0,T)$ are marked with ○.

zero, getting simply $u'\,dt$ times the value taken by (3-59) with $u = 0$, because of the smallness of $dt$. Integrating this result from 0 to $\infty$ with respect to $u'$, we obtain $\rho\,dt$ for the probability of an upward zero crossing in any interval of length $dt$.

Since the number of times $u(t)$ decreases through zero in *any* interval must be very nearly equal to the number of times it increases through zero, the probability of a downward zero crossing is also $\rho\,dt$, that is, the probability of a zero in an interval of length $dt$ is $2\rho\,dt$. Thus, dividing an interval of length $T$ into increments $dt$, we see that, on account of the stationarity of $u(t)$, the expected number of zeros during the time $T$ is the integral of $2\rho\,dt$, namely,

$$\begin{aligned} 2\rho T &= 2T\sqrt{\frac{\int_{-\infty}^{\infty} f^2\Psi(f)\,df}{\int_{-\infty}^{\infty}\Psi(f)\,df}} \\ &= \frac{T}{\pi}\sqrt{-\frac{\psi''(0)}{\psi(0)}}, \end{aligned} \tag{3-60}$$

the expected number per unit time being[1] $2\rho$.

[1] Rice, *op. cit.*, eq. (3.3-11).

**Problem 3-21.** Show that, if $u(t)$ is an ergodic, zero-mean, gaussian random process with rms spectral width $\rho$ and rms value $\sigma$, the average number of times per second that $u(t) = c$ is $2\rho \exp(-\frac{1}{2}c^2/\sigma^2)$.

**Problem 3-22.** Show that a gaussian random process whose spectral density is constant between $f = \pm\frac{1}{2}B$ and is zero otherwise has $B/\sqrt{3}$ zeros per second on the average.

**Problem 3-23.** Show that for the noise of Prob. 3-22 the average number of maxima per second is $\sqrt{3/20}\,B$, and explain why this exceeds half of $B/\sqrt{3}$. What is the average number of points of inflection per unit time?

## □ OPTIMUM LINEAR FILTERING†

If $x_1$ and $x_2$ are jointly normal, with known means and covariance matrix, then (Prob. 1-8) for a given $x_1$ the best estimate $v$ of $x_2$ is a linear function of $x_1$, namely, the conditional expectation $\mathrm{E}\{x_2|x_1\}$ given by (1-14)—best in the sense that it minimizes the mean value of any nondecreasing function of the error $|x_2 - v|$, such as the mean-squared error $\mathrm{E}\{(x_2 - v)^2\}$. Similarly, if a vector $\mathbf{u}$ and a number $w$ are jointly normal with known means and covariance matrix, the best estimate of $w$ for a given $\mathbf{u}$ will be a linear function of $\mathbf{u}$ (i.e., of its components). Thus, if $u(t)$ and $w(t)$ are jointly normal zero-mean ergodic random processes with known autocorrelation functions $\psi_u(\tau)$ and $\psi_w(\tau)$, respectively, and crosscorrelation function (see Prob. 3-3) $\psi_{uw}(\tau) = \mathrm{E}\{u(t)w(t+\tau)\}$, then the best estimate of $w(t)$, given the values of $u(\tau)$ for $-\infty < \tau \leq t$, for example, is a linear function of these values, say

$$v(t) = \int_0^\infty u(t-\tau)g(\tau)\,d\tau, \tag{3-61}$$

that is, the output of a certain linear filter with $u(t)$ as input. When $u(t)$ and $w(t)$ are not jointly normal, the best estimate of $w(t)$ is generally not of this form, but it is sometimes of interest to find the linear filter $g(t)$ that minimizes the mean-squared error. This filter will be the best possible estimator of $w(t)$ when $u(t)$ and $w(t)$ are jointly gaussian.

When $u(t)$ equals $w(t)$ plus independent random noise, we have the "smoothing" problem of separating $w(t)$ from the noise on the basis of earlier values of $u(t)$ alone. If $u(t) = w(t - T)$, we have the problem of predicting or "extrapolating" $u(t)$ a time $T$ into the future. If $u(t)$ is $w(t + T)$ plus noise, we have the "interpolation" problem.[1] Here we

† May be omitted on the first reading.

[1] Norbert Wiener, "Extrapolation, Interpolation, and Smoothing of Stationary Time Series," John Wiley & Sons, Inc., New York, 1949; and Chapman & Hall, Ltd., London, 1949.

obtain a more accurate value of $w(t)$ than in the first case by making use of later values of $u(t)$. Among the other problems that take the foregoing form is that of determining $w'(t)$ from the set of all earlier values of $u(t)$, the sum of $w(t)$ plus noise.

With $g(t)$, the impulse response of the filter, taking the value zero for negative $t$, we can express (3-61) as the convolution $v = u \star g$, and we can write the mean-squared error as

$$\begin{aligned}
\mathrm{E}\{[w(t) - v(t)]^2\} &= \mathrm{E}\{(w - u \star g)^2\} \\
&= \mathrm{E}\{w^2\} - 2\mathrm{E}\{w(u \star g)\} + \mathrm{E}\{(u \star g)^2\} \\
&= \psi_w(0) - 2\mathrm{E}\left\{w(t) \int_0^\infty u(t-\tau)g(\tau)\,d\tau\right\} \\
&\qquad + \mathrm{E}\left\{\int_0^\infty u(t-\tau)g(\tau)\,d\tau \int_0^\infty u(t-v)g(v)\,dv\right\} \\
&= \psi_w(0) - 2\int_0^\infty \psi_{uw}(\tau)g(\tau)\,d\tau \\
&\qquad + \iint_0^\infty \psi_u(\tau - v)g(\tau)g(v)\,d\tau\,dv. \qquad (3\text{-}62)
\end{aligned}$$

To find the filter impulse response $g(t)$ that minimizes this mean-squared difference between the filter output $v(t)$ and the desired value $w(t)$, we regard the integrals in (3-62) as summations of terms containing unknown numbers called $g(\tau)$ for many different values of $\tau$, and we set the derivative of (3-62) with respect to $g(\tau)$ equal to zero, getting

$$-2\psi_{uw}(\tau) + 2\int_0^\infty \psi_u(\tau - v)g(v)\,dv = 0.$$

Hence the optimum filter is the one whose impulse response satisfies the integral equation

$$\int_0^\infty \psi_u(\tau - v)g(v)\,dv = \psi_{uw}(\tau). \qquad (3\text{-}63)$$

We shall not go further into this matter,[1] but instead, shall turn now to the representation and statistics of narrow-band waveforms, which find widespread use in communication systems. Generally, the desired receiver output $w(t)$ is the waveform that modulated the transmitted signal, but it and the received signal $u(t)$, which may be the transmitted signal plus noise, are seldom jointly gaussian, and nonlinear (or time-varying) processing of $u(t)$ by a demodulator is required.

[1] See Y. W. Lee, "Statistical Theory of Communication," chaps. 14–16, John Wiley & Sons, Inc., New York, 1960; J. H. Laning and R. H. Battin, "Random Processes in Automatic Control," chap. 7, McGraw-Hill Book Company, New York, 1956; Middleton, *op. cit.*, chap. 16.

**Problem 3-24.** Express $\psi_{uw}(\tau)$ in terms of $\psi_u(\tau)$ and $\psi_w(\tau)$ for the cases of smoothing, extrapolation, and interpolation, and of smoothing the derivative.

**Problem 3-25.** What form would (3-62) take if our estimate of $w(t)$ were based on the values of $u(\tau)$ for *all* $\tau$? (Such an estimate is possible, of course, only with a long delay that permits all relevant parts of $u(\tau)$ to be observed.) Show that the Fourier transform of the $g(t)$ that satisfies the resulting integral equation is $G(f) = \Psi_{uw}(f)/\Psi_u(f)$ (see Prob. 3-3), and apply this result to the cases studied in Prob. 3-24.

**Problem 3-26.** Show that the minimum mean-squared error in the general case described in Prob. 3-25 is

$$\mathrm{E}\,\{(w - u \star g)^2\} = \int_{-\infty}^{\infty} \left[ \Psi_w(f) - \frac{|\Psi_{uw}(f)|^2}{\Psi_u(f)} \right] df.$$

**Problem 3-27.** Show that, if there is a fixed $h(t)$ such that $u(t)$ is $w(t) \star h(t)$ plus noise statistically independent of $w(t)$, then $\psi_{uw}(\tau) = \psi_w(\tau) \star h(\tau)$, and apply this result to the case of $h(t) = -\delta'(t)$.

**Problem 3-28.** By a derivation paralleling that of (3-10), show that $\Psi_{uw}(f) = \lim_{T\to\infty} \mathrm{E}\,\{U_T^*(f)W_T(f)\}/T$, and use Schwarz's inequality[1] to show that the integrand in Prob. 3-26 cannot be negative, that is, $|\Psi_{uw}(f)|^2 \leq \Psi_u(f)\Psi_w(f)$ for every $u$, $w$, and $f$.

[1] Since $\iint_{-\infty}^{\infty} |Q(s)R^*(t) - Q^*(t)R(s)|^2 \, ds\, dt \geq 0$ for any functions $Q(t)$ and $R(t)$, with equality if and only if their ratio is a constant, on expanding we have Schwarz's inequality $\left|\int_{-\infty}^{\infty} Q(t)R(t)\, dt\right|^2 \leq \int_{-\infty}^{\infty} |Q(t)|^2\, dt \int_{-\infty}^{\infty} |R(t)|^2\, dt.$

# chapter 4 NARROW–BAND RANDOM PROCESSES

A random process $u(t)$ whose power spectral density $\Psi_u(f)$ vanishes, insofar as positive frequencies are concerned, outside a band whose width is small compared to the frequencies in it is called *narrow-band*. Its spectrum therefore consists of two widely separated parts, mirror images in the line $f = 0$, as in the cases of (3-35) and (3-41).

If $u_T(t)$, (3-4), is expressed as a Fourier series, we see by (3-10) that, as $T \to \infty$, those terms become negligible whose frequencies lie outside the band occupied by $u(t)$, for the portion of the total power

$$\mathrm{E}\,\{u^2(t)\} = \psi_u(0) = \int_{-\infty}^{\infty} \Psi_u(f)\,df$$

coming from outside this band is zero. Denoting by $F$ an arbitrary frequency within the positive-frequency band occupied by $u(t)$, then, we see that $u(t)$ can be regarded as made up of Fourier components whose frequencies are relatively close to either $F$ or $-F$.

Thus, combining negative with positive frequencies, we conclude that a narrow-band waveform can be regarded as the sum of many sinusoids of nearly equal frequencies. Their phase relationship, therefore, changes significantly only after many cycles of oscillation—in a time of the order of the reciprocal of the bandwidth—and their sum appears to be a sinusoid of frequency $F$ whose amplitude $a$ and phase $\phi$ change very little from one cycle to the next. Thus we can express it as

$$u(t) = a(t) \cos [2\pi Ft + \phi(t)]. \tag{4-1}$$

For different choices of the reference frequency $F$ within the band, $\phi(t)$ will be different, but in *any* case it will change very little from one cycle of the reference frequency to the next. Since the effects of small, rapid changes in $a(t)$ and in $\phi(t)$ in (4-1) are indistinguishable, $u(t)$ does not determine $a(t)$ and $\phi(t)$ completely. However, the larger the ratio of $F$ to the bandwidth, the more slowly they change and the more precisely they can be determined by observing the oscillation of $u(t)$. In this chapter we shall study their statistics.

We shall suppose $u(t)$ to be ergodic. On account of the ambiguity of phase angles, $\phi(t)$ contains an arbitrary multiple of $2\pi$. Because of the ergodicity of $u(t)$, $\phi(t)$ must contain, in addition, a constant like the $\theta$ in (3-33) or (3-36), which is distributed uniformly between 0 and $2\pi$, so that all translates of (4-1) will be equally probable. Consequently,

$$0 = \mathrm{E}\{\cos \phi(t)\} = \mathrm{E}\{\sin \phi(t)\} = \mathrm{E}\{\sin \phi(t) \cos \phi(t)\} \tag{4-2}$$

and

$$\tfrac{1}{2} = \mathrm{E}\{\cos^2 \phi(t)\} = \mathrm{E}\{\sin^2 \phi(t)\}. \tag{4-3}$$

Since the sum $\phi(t) + \phi(t + \tau)$ contains twice this uniformly distributed constant, we have

$$0 = \mathrm{E}\{\cos [\phi(t) + \phi(t + \tau)]\} = \mathrm{E}\{\sin [\phi(t) + \phi(t + \tau)]\}. \tag{4-4}$$

Using the formula for the product of two cosines, and (4-4), we find that the correlation function of (4-1) is

$$\begin{aligned}\psi_u(\tau) &= \mathrm{E}\{u(t)u(t + \tau)\} \\ &= \tfrac{1}{2}\mathrm{E}\{a(t)a(t + \tau) \cos [2\pi F\tau + \phi(t + \tau) - \phi(t)]\}. \end{aligned} \tag{4-5}$$

Proceeding as in (3-39) and writing the cosine as the sum of two imaginary exponentials, we get

$$\psi_u(\tau) = \tfrac{1}{2}z(\tau)e^{2\pi iF\tau} + \tfrac{1}{2}z^*(\tau)e^{-2\pi iF\tau}, \tag{4-6}$$

where

$$z(\tau) = \mathrm{E}\{a(t)a(t + \tau)e^{i\phi(t+\tau)-i\phi(t)}\}. \tag{4-7}$$

The Fourier transform of (4-6) is the power spectral density of $u(t)$,

$$\Psi_u(f) = \tfrac{1}{2}Z(f - F) + \tfrac{1}{2}Z(-f - F), \tag{4-8}$$

where $Z(f)$ is the Fourier transform of (4-7). This is the generalization of (3-41) to a random process whose amplitude $a(t)$, as well as its instantaneous frequency $F + (2\pi)^{-1}\,d\phi(t)/dt$, varies. With $F$ very large, the spectrum of $u(t)$ consists of two distinct parts, one around $F$ and the other around $-F$, corresponding to the respective terms of (4-8). Thus for $f \geq 0$ the second term vanishes, and we have

$$\Psi_u(f) = \tfrac{1}{2}Z(f - F). \tag{4-9}$$

**Problem 4-1.** Show that the centroid [defined as in the first line of (3-49)] of the positive-frequency part of the spectrum of any narrow-band waveform (4-1), i.e., the centroid of the first term on the right-hand side of (4-8), is

$F + \mu$ with $\mu = \mathrm{E}\{a^2\phi'\}/2\pi\,\mathrm{E}\{a^2\}$, and show that its rms bandwidth [defined as in the first line of (3-48)] is

$$\sqrt{\frac{\mathrm{E}\{a'^2 + a^2\phi'^2\}}{4\pi^2\,\mathrm{E}\{a^2\}} - \mu^2},$$

where $a' = da/dt$ and $\phi' = d\phi/dt$. [HINT: Express (4-7) and its first two derivatives evaluated at $\tau = 0$ in terms of its Fourier transform. Replace $t$ by $t - \tau$ before differentiating (4-7) the second time, as in (3-19) (see Prob. 3-17).] Being the sum of the squares of the radial and tangential components of the velocity, $a'^2 + a^2\phi'^2$ is the square of the speed of motion of the phasor representing the narrow-band waveform.

**Orthogonal Components.** Using the formula for the cosine of a sum, we can write (4-1) as

$$u(t) = x(t)\cos 2\pi Ft - y(t)\sin 2\pi Ft, \tag{4-10}$$

where

$$x(t) = a(t)\cos\phi(t), \qquad y(t) = a(t)\sin\phi(t). \tag{4-11}$$

Since $u = \mathrm{Re}\{ae^{i\phi}e^{2\pi iFt}\} = \mathrm{Re}\{(x + iy)e^{2\pi iFt}\}$, we see that

$$x + iy = ae^{i\phi}$$

is the phasor representing $u(t)$ in terms of the reference frequency $F$.

Using (4-4), we find that the correlation function of $x(t)$, the magnitude of the component of $u(t)$ in phase with $\cos 2\pi Ft$, is

$$\begin{aligned}\psi_x(\tau) &= \mathrm{E}\{a(t)a(t+\tau)\cos\phi(t)\cos\phi(t+\tau)\}\\ &= \tfrac{1}{2}\mathrm{E}\{a(t)a(t+\tau)\cos[\phi(t+\tau) - \phi(t)]\}\\ &= \tfrac{1}{2}z(\tau) + \tfrac{1}{2}z^*(\tau) = \mathrm{Re}\{z(\tau)\},\end{aligned} \tag{4-12}$$

which is the same as (4-5) and (4-6) with $F = 0$. Thus the power spectral density of $x(t)$ is

$$\Psi_x(f) = \tfrac{1}{2}Z(f) + \tfrac{1}{2}Z(-f), \tag{4-13}$$

which may be recognized as the *even part* of the function $Z(f)$. Using (4-4) and the formula for the product of two sines, we find that the correlation function of $y(t)$, the magnitude of the component of $u(t)$ in phase with $-\sin 2\pi Ft$, is (4-12), like that of $x(t)$. Its power spectral density, too, is therefore the same as that of $x(t)$,

$$\Psi_y(f) = \tfrac{1}{2}Z(f) + \tfrac{1}{2}Z(-f). \tag{4-14}$$

Notice that, when the spectrum of $u(t)$ is symmetric about the reference frequency $F$, that is, when $Z(f) = Z(-f)$, the power spectral densities of $x(t)$ and of $y(t)$ are simply $Z(f)$, and, since their Fourier transform $z(\tau)$ is then real, (4-6) takes the form

$$\psi_u(\tau) = z(\tau)\cos 2\pi F\tau. \tag{4-15}$$

**Moments of $x$, $y$, $x'$ and $y'$.** From (4-1), (4-2), and (4-11), we see that $u(t)$, $x(t)$, and $y(t)$ all have mean value zero. Since

$$\mathrm{E}\,\{dx(t)/dt\} = (d/dt)\,\mathrm{E}\,\{x(t)\}$$

and $\mathrm{E}\,\{dy(t)/dt\} = (d/dt)\,\mathrm{E}\,\{y(t)\}$, we see that $x'(t)$ and $y'(t)$ also have mean value zero.

From (4-2) and (4-11) we see that $\mathrm{E}\,\{x(t)y(t)\} = 0$, that is, that $x(t)$ and $y(t)$ at the same instant are uncorrelated. Since

$$\mathrm{E}\,\{x(t)x'(t)\} = \frac{1}{2}\left(\frac{d}{dt}\right)\mathrm{E}\,\{x^2(t)\},$$

and since the ergodicity of $u(t)$ implies the ergodicity of $x(t)$, $\mathrm{E}\,\{x^2(t)\}$ is a constant, and $x(t)$ and $x'(t)$ at the same instant are uncorrelated. Likewise, $\mathrm{E}\,\{y(t)y'(t)\} = 0$.

Setting $\tau = 0$ in (4-6), (4-7), and (4-12), which is $\psi_y(\tau)$ as well as $\psi_x(\tau)$, we see that $u(t)$, $x(t)$, and $y(t)$ all have the same mean-squared value, $\frac{1}{2}\,\mathrm{E}\,\{a^2(t)\}$, which we shall denote by $\sigma^2$; that is,

$$\mathrm{E}\,\{x^2(t)\} = \mathrm{E}\,\{y^2(t)\} = \mathrm{E}\,\{u^2(t)\} = \tfrac{1}{2}\,\mathrm{E}\,\{a^2(t)\} = \sigma^2. \qquad (4\text{-}16)$$

By (3-20), the power spectral density of $x'(t)$, like that of $y'(t)$, is $4\pi^2 f^2$ times (4-13); hence

$$\begin{aligned}\mathrm{E}\,\{x'^2(t)\} = \mathrm{E}\,\{y'^2(t)\} &= 2\pi^2 \int_{-\infty}^{\infty} f^2[Z(f) + Z(-f)]\,df \\ &= 4\pi^2 \int_{-\infty}^{\infty} f^2 Z(f)\,df \\ &= 4\pi^2\rho^2\sigma^2, \end{aligned} \qquad (4\text{-}17)$$

where $\rho$ is the rms width of the spectra of $x(t)$ and $u(t)$, defined as in (3-42) and (3-48), respectively. Here we have made use of the fact that $\sigma^2 = \int_{-\infty}^{\infty} \Psi_x(f)\,df = \int_{-\infty}^{\infty} Z(f)\,df$.

Differentiating (4-11), we get

$$x' = a'\cos\phi - a\phi'\sin\phi, \qquad y' = a'\sin\phi + a\phi'\cos\phi, \qquad (4\text{-}18)$$

and, making use of (4-2) and (4-3), we find that $\mathrm{E}\,\{x'(t)y'(t)\} = 0$. Thus four of the six covariances among $x(t)$, $y(t)$, $x'(t)$, and $y'(t)$ always vanish. The other two, $\mathrm{E}\,\{x(t)y'(t)\}$ and $\mathrm{E}\,\{x'(t)y(t)\}$, however, need not be zero; in fact,

$$\begin{aligned} E\,\{xy'\} &= \mathrm{E}\,\{a\cos\phi\,(a'\sin\phi + a\phi'\cos\phi)\} \\ &= \tfrac{1}{2}\,\mathrm{E}\,\{a^2\phi'\}. \end{aligned} \qquad (4\text{-}19)$$

The derivative of (4-7) with respect to $\tau$ at $\tau = 0$,

$$\begin{aligned} z'(0) &= \tfrac{1}{2}\,\mathrm{E}\,\{ia^2\phi' + aa'\} \\ &= \tfrac{1}{2}i\,\mathrm{E}\,\{a^2\phi'\}, \end{aligned} \qquad (4\text{-}20)$$

can also be written as

$$\begin{aligned} z'(0) &= 2\pi i \int_{-\infty}^{\infty} fZ(f)\, df \\ &= 2\pi i \int_{-\infty}^{\infty} (f - F)Z(f - F)\, df \\ &= 2\pi i \int_{0}^{\infty} (f - F)\Psi_u(f)\, df \end{aligned} \tag{4-21}$$

by differentiating $z(\tau) = \int_{-\infty}^{\infty} Z(f)e^{2\pi i f\tau}\, df$ and using (4-9). Comparing (4-19), (4-20), and (4-21), we see that, when the centroid of the positive-frequency part of the spectrum of $u(t)$ is taken as the reference frequency $F$, $\mathrm{E}\,\{x(t)y'(t)\} = 0$. Likewise, $\mathrm{E}\,\{x'(t)y(t)\} = -\frac{1}{2}\,\mathrm{E}\,\{a^2(t)\phi'(t)\}$ vanishes when

$$F = \frac{\int_0^{\infty} f\Psi_u(f)\, df}{\int_0^{\infty} \Psi_u(f)\, df}. \tag{4-22}$$

Thus, with (4-22) as reference frequency, of the four quantities $x(t)$, $y(t)$, $x'(t)$, $y'(t)$ for the same $t$, no pair is correlated.

In general, however, the values of these quantities at *different* times *are* correlated. For example,

$$\begin{aligned} \mathrm{E}\,\{x(t)y(t + \tau)\} &= \tfrac{1}{2}\,\mathrm{E}\,\{a(t)a(t + \tau) \sin [\phi(t + \tau) - \phi(t)]\} \\ &= \tfrac{1}{2}iz^*(\tau) - \tfrac{1}{2}iz(\tau) \\ &= \mathrm{Im}\,\{z(\tau)\}. \end{aligned} \tag{4-23}$$

This vanishes for all $\tau$ only when $z(\tau)$ is real for all $\tau$, that is, when $\psi_u(\tau)$ is of the form (4-15) and $Z(f)$ is an even function. Thus $x(s)$ and $y(t)$ are uncorrelated for all $s$ and $t$ if and only if the positive-frequency part of the spectrum of $u(t)$ is symmetric about $f = F$.

## □ NARROW - BAND GAUSSIAN NOISE

Whenever $u(t)$ is an ergodic gaussian random process, $x(t)$ and $y(t)$ are jointly ergodic gaussian random processes, for $x(t)$ and $y(t)$ do not change significantly in a time $1/F$, and, by (4-10), $x(t) = u(t)$ whenever

$$\cos 2\pi Ft = 1,$$

and $y(t) = u(t)$ whenever $\sin 2\pi Ft = -1$. That is, $x(t_1)$, $x(t_2)$, . . . , $x(t_m)$, $y(t_1)$, $y(t_2)$, . . . , $y(t_m)$ obey a $2m$-variate normal distribution for any positive integer $m$, with moments depending only on time differences.

In particular, $x(t)$ and $y(t)$ at the same instant, being jointly normal with zero mean and zero covariance and having equal variances (4-16), obey the circular normal distribution (1-17) with probability density function

$$p(x,y|\sigma) = \frac{1}{2\pi\sigma^2} \exp\left(-\frac{x^2+y^2}{2\sigma^2}\right). \tag{4-24}$$

From (4-11) it follows as in (1-18) that the phase $\phi$, if reduced by a multiple of $2\pi$ so as to lie between 0 and $2\pi$, is distributed uniformly over this interval, as required by the ergodicity of $u(t)$, and the amplitude $a$ obeys the Rayleigh distribution (1-20), being independent of the value of $\phi$ at the same instant.

**Problem 4-2.** Show that the ratio $2/\sqrt{\pi}$ of the rms value to the mean of the Rayleigh distribution is the factor by which an a-c meter reading must be multiplied when used to measure zero-mean gaussian noise if the meter contains a linear rectifier so that it measures the average of its positive inputs but is calibrated in terms of the rms value of a sinusoid.

**Sinusoid Plus Gaussian Noise.** We shall now suppose that the narrow-band random process (4-1) is the sum of a sinusoid of amplitude $A$ plus gaussian noise described by (4-24). Since the reference frequency $F$ is arbitrary, we may take it to be the frequency of the sinusoid, and we can choose the origin of time so that the latter is $A \cos 2\pi Ft$.

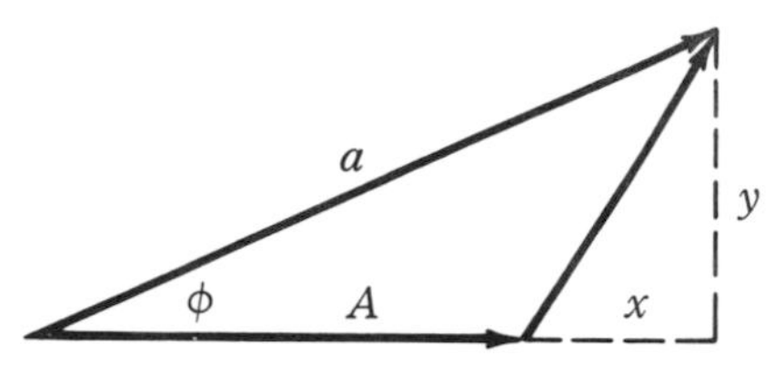


*Figure 4-1.* Phasor diagram of the sum of narrow-band noise plus a sinusoid of amplitude $A$ and phase 0.

Adding this sinusoid to $x(t) \cos 2\pi Ft - y(t) \sin 2\pi Ft$ and equating their sum to $a(t) \cos [2\pi Ft + \phi(t)]$, we see (Fig. 4-1) that

$$x = a \cos \phi - A, \qquad y = a \sin \phi. \tag{4-25}$$

Substituting (4-25) into (4-24) and multiplying by the jacobian (1-19), we obtain the joint probability density function of $a$ and $\phi$,

$$p(a,\phi|A,\sigma) = \frac{a}{2\pi\sigma^2} \exp\left(-\frac{A^2 - 2Aa\cos\phi + a^2}{2\sigma^2}\right). \tag{4-26}$$

Since (4-26) cannot be written as the product of a function of $\phi$ times a function of $a$, these two quantities are not statistically independent for $A \neq 0$.

To find the distribution of $a$ alone we integrate (4-26) with respect to $\phi$ from 0 to $2\pi$, making use of the relation

$$I_m(\xi) = \frac{1}{2\pi} \int_0^{2\pi} e^{\xi\cos\phi} \cos m\phi \, d\phi \tag{4-27}$$

for the modified Bessel function of the first kind, of integral order $m$. Evidently, $I_0(0) = 1$ and $I_m(0) = 0$ for $m = 1, 2, \ldots$. From (4-27) we see that $I_m(\xi)$ increases monotonically for $\xi \geq 0$ and that for $\xi \gg |m| + 1$ we can use the approximation

$$I_m(\xi) = \frac{e^\xi}{\sqrt{2\pi\xi}}. \tag{4-28}$$

$I_m(\xi)$ can be expressed in terms of ordinary Bessel functions of the first kind as $J_m(i\xi)/i^m$, and it is sometimes[1] tabulated under this heading.

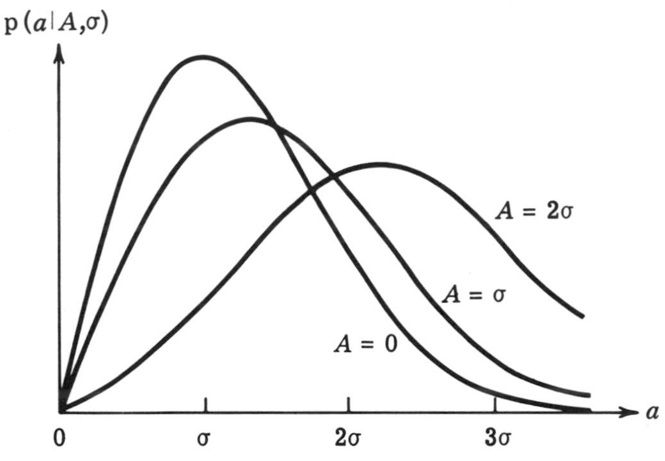


*Figure 4-2.* The probability density function $p(a|A,\sigma)$ of the amplitude $a$ of the sum of a sinusoid of amplitude $A$ plus narrow-band gaussian noise of rms value $\sigma$ for $A = 0$, $\sigma$, and $2\sigma$.

Using (4-27) to integrate (4-26) with respect to $\phi$, we thus find that the probability density function of $a$ is

$$p(a|A,\sigma) = \frac{a}{\sigma^2} I_0\left(\frac{Aa}{\sigma^2}\right) \exp\left(-\frac{A^2 + a^2}{2\sigma^2}\right), \tag{4-29}$$

which is shown for several values of $A$ in Fig. 4-2. With $A = 0$, (4-29) of course becomes the Rayleigh distribution (1-20). For $A \gg \sigma$, as we can see from Fig. 4-1, $a$ is approximately $A + x$ and is therefore normally distributed with mean $A$ and variance $\sigma^2$. The substitution of (4-28) into (4-29) for large $A$ leads to the same conclusion.

If we multiply (4-26) by $a^n \cos m\phi$ and integrate with respect to $\phi$ from 0 to $2\pi$ by means of (4-27) and with respect to $a$ from 0 to $\infty$ by

[1] E. Jahnke and F. Emde, "Funktionentafeln mit Formeln und Kurven," B. G. Teubner Verlagsgesellschaft, mbH, Leipzig, 1933; 6th ed., McGraw-Hill Book Company, New York, 1960. Chapter X of this book presents graphs of the confluent hypergeometric function.

means of a table of definite integrals,[1] we find that for any $n > -2$ and any integer $m$ the mean value of $a^n \cos m\phi$ is

$$\mathrm{E}\,\{a^n \cos m\phi\} = \left(\frac{m+n}{2}\right)! \frac{A^m}{m!} (\sqrt{2}\,\sigma)^{n-m} {}_1F_1\left(\frac{m-n}{2};m+1;-\frac{A^2}{2\sigma^2}\right), \tag{4-30}$$

where

$${}_1F_1(\alpha;\gamma;-r) = 1 - \frac{\alpha}{\gamma}\frac{r}{1!} + \frac{\alpha}{\gamma}\frac{\alpha+1}{\gamma+1}\frac{r^2}{2!} - \frac{\alpha}{\gamma}\frac{\alpha+1}{\gamma+1}\frac{\alpha+2}{\gamma+2}\frac{r^3}{3!} + \cdots \tag{4-31}$$

is the *confluent hypergeometric function*. When $\alpha$ is a nonpositive integer, (4-31) is a polynomial in $r$, and for other special values of $\alpha$ and $\gamma$ it can be expressed in terms of familiar functions.[2] This function satisfies Kummer's first relation,

$${}_1F_1(\alpha;\gamma;-r) = e^{-r}{}_1F_1(\gamma-\alpha;\gamma;r), \tag{4-32}$$

and for $r$ large compared to $|\alpha| + 1$ and $|\gamma - \alpha|$ is approximately

$${}_1F_1(\alpha;\gamma;-r) = \frac{(\gamma-1)!}{(\gamma-\alpha-1)!} r^{-\alpha}. \tag{4-33}$$

Note that $n$ in (4-30) need not be an integer and that by $s!$ we mean $\Gamma(s+1)$, (1-7), when $s$ is not an integer.

In Chap. 5 we shall make full use of (4-30) along with the mean value

$$\mathrm{E}\,\{a^n \sin m\phi\} = 0, \tag{4-34}$$

which vanishes because (4-26) is an even function of $\phi$. Here, setting $m = 0$, we get the $n$th moment of the distribution (4-29) of $a$,

$$\mathrm{E}\,\{a^n\} = (\tfrac{1}{2}n)!(\sqrt{2}\,\sigma)^n {}_1F_1(-\tfrac{1}{2}n;1;-A^2/2\sigma^2). \tag{4-35}$$

**Problem 4-3.** From the relation $a^2 = (A+x)^2 + y^2$ and the even moments (1-4) of the normal distribution of $x$ and $y$, show that $\mathrm{E}\,\{a^2\} = A^2 + 2\sigma^2$ and $\mathrm{E}\,\{a^4\} = A^4 + 8A^2\sigma^2 + 8\sigma^4$, and find $\mathrm{E}\,\{a^6\}$ and $\mathrm{E}\,\{a^8\}$. Verify that these

[1] E. Madelung, "Die Mathematischen Hilfsmittel des Physikers," 6th ed., p. 110, Springer-Verlag OHG, Berlin, 1957; Bateman Manuscript Project, "Tables of Integral Transforms," vol. 2, p. 30 No. 14, McGraw-Hill Book Company, New York, 1954, or vol. 1, p. 186 No. 35.

[2] Middleton, *op. cit.*, app. 1.2, pp. 1073–1076; S. O. Rice, *op. cit.*, app. 4B; M. Abramowitz and I. A. Stegun (eds.), "Handbook of Mathematical Functions," Applied Mathematics Series, vol. 55, chap. 13, pp. 503–535, National Bureau of Standards, Washington, 1964.

results agree with (4-35) and that, for $A = 0$, (4-35) is in agreement with the results of Prob. 1-10.

**Problem 4-4.** (*a*) From (4-26) obtain the probability density function of $\phi$,

$$(2\pi)^{-1}e^{-r} + \sqrt{r/\pi}\, e^{-r \sin^2\phi} \cos\phi \operatorname{erf}(\sqrt{2r}\cos\phi),$$

where $r = A^2/2\sigma^2$ is the ratio of the power of the sinusoid to that of the noise.

(*b*) With the help of Fig. 4-1, describe the distribution of $\phi$ for $A \gg \sigma$. Here it is better to restrict $\phi$ to the range $(-\pi,\pi)$ than to $(0,2\pi)$.

**Problem 4-5.** Substituting (4-33) into (4-30) gives $\mathrm{E}\{a^n \cos m\phi\} = A^n$ because, in the absence of noise, $a = A$ and $\phi = 0$. By using three terms of the binomial series for $a^n$ with $a^2 = (A + x)^2 + y^2$ and two terms of the power series for $\cos m\phi$ with $\phi \cong y/A$, obtain the two-term approximation $\mathrm{E}\{a^n \cos m\phi\} = A^n + \frac{1}{2}(n^2 - m^2)A^{n-2}\sigma^2$ for $A \gg \sigma$.

**Problem 4-6.** Show that the first two moments of $\cos\phi$ are

$$\mathrm{E}\{\cos\phi\} = \tfrac{1}{2}\sqrt{\pi r}\, e^{-r/2}[I_0(\tfrac{1}{2}r) + I_1(\tfrac{1}{2}r)]$$

and

$$\mathrm{E}\{\cos^2\phi\} = 1 - (1 - e^{-r})/2r,$$

where $r = A^2/2\sigma^2$.

**Problem 4-7.** Show that $\mathrm{E}\{(\cos\phi)/a\} = (1 - e^{-r})/A$.

**Problem 4-8.** A phase meter consists of a flip-flop with an output of either 0 or 360 volts and a voltmeter which reads its average output, the flip-flop output being flipped to 360 volts by each upward zero-crossing of a reference waveform $\cos 2\pi Ft$ and flopped to 0 by each upward zero-crossing of another input, which is the sum of a sinusoid $A \cos(2\pi Ft + \theta)$ plus narrow-band gaussian noise of rms value $\sigma$. Use (4-24) to show that for $\theta = 0$ or 180° the average meter reading will be 180, while for $\theta = 90°$ or 270° it will be $90 + 180 \operatorname{erf}(-A/\sigma)$ or $270 - 180 \operatorname{erf}(-A/\sigma)$, respectively.

## □ THE DISTRIBUTION OF THE RATES OF CHANGE

Returning to the case where $u(t)$, (4-1) and (4-10), is an ergodic gaussian random process and does not include a sinusoid, we note that $x(t)$, $y(t)$, $x'(t)$, and $y'(t)$ for any single value of $t$ are jointly normal, since the latter two are simply the limits of differences of the former two. We suppose that the centroid of the spectrum of the noise is chosen as the reference frequency $F$ so that these four quantities will be uncorrelated and hence statistically independent of one another.

Since their means are zero and their variances are (4-16) and (4-17), their joint probability density function is the product of (4-24) times a similar density function of $x'$ and $y'$,

$$p(x,y,x',y'|\rho,\sigma) = \frac{\exp\left(-\frac{x^2+y^2}{2\sigma^2}\right)\exp\left(-\frac{x'^2+y'^2}{8\pi^2\rho^2\sigma^2}\right)}{16\pi^4\rho^2\sigma^4}. \quad (4\text{-}36)$$

Here we see that the velocity $x' + iy'$ of the phasor $x + iy$ representing $u(t)$ is statistically independent of the phasor itself and obeys a circular normal distribution. The direction of motion is therefore uniformly distributed, and the speed of motion $s = \sqrt{x'^2 + y'^2}$ is independent of it and is Rayleigh-distributed.

Substituting (4-11) and (4-18) into (4-36) and multiplying by the jacobian of $x$, $y$, $x'$, $y'$ with respect to $a$, $\phi$, $a'$, $\phi'$ [that is, the determinant whose elements are the 16 first partial derivatives, similar to (1-19), whose value turns out to be $a^2$], we get the joint probability density function of $a$, $\phi$, $a'$, and $\phi'$,

$$p(a,\phi,a',\phi'|\rho,\sigma) = \frac{a^2}{16\pi^4\rho^2\sigma^4}\exp\left[-\frac{(4\pi^2\rho^2+\phi'^2)a^2+a'^2}{8\pi^2\rho^2\sigma^2}\right], \quad (4\text{-}37)$$

since $dx\,dy\,dx'\,dy' = a^2\,da\,d\phi\,da'\,d\phi'$, just as $dx\,dy = a\,da\,d\phi$.

Since $\phi$ does not appear on the right-hand side of (4-37), it follows that $\phi$ is statistically independent of $a$, $a'$, and $\phi'$ and, as we already know, is distributed uniformly between 0 and $2\pi$. In fact, (4-37) can be written as the product of two factors, one depending only on $a$ and $\phi'$, and the other depending only on $a'$. Hence $a'$ is statistically independent of $a$, $\phi$, and $\phi'$, and as the form of the second factor shows, $a'$ is normally distributed with mean zero and variance $4\pi^2\rho^2\sigma^2$. The first factor gives us the joint probability density function of $a$ and $\phi'$,

$$p(a,\phi'|\rho,\sigma) = \frac{a^2}{(2\pi)^{\frac{3}{2}}\rho\sigma^3}\exp\left(-\frac{4\pi^2\rho^2+\phi'^2}{8\pi^2\rho^2\sigma^2}a^2\right). \quad (4\text{-}38)$$

If we integrate (4-38) with respect to $\phi'$ from $-\infty$ to $\infty$, we, of course, obtain the Rayleigh distribution (1-20) of $a$. Integrating (4-38) with respect to $a$ from 0 to $\infty$, we find that the probability density function of the instantaneous frequency deviation $\phi'$ (in radians per second) of narrow-band gaussian noise of spectral width $\rho$ is

$$p(\phi'|\rho) = \frac{2\pi^2\rho^2}{(4\pi^2\rho^2+\phi'^2)^{\frac{3}{2}}}. \quad (4\text{-}39)$$

This is a bell-shaped distribution with mean value E $\{\phi'\} = 0$ and mean absolute value E $\{|\phi'|\} = 2\pi\rho$ radians per second; however, E $\{\phi'^2\} = \infty$.

As phase reference we have been using $\cos 2\pi Ft$, but, because of the ergodicity of $u(t)$, we would have observed the same behavior of $x(t)$, $y(t)$, $a(t)$, $\phi(t)$, and their derivatives if we had used $\cos(2\pi Ft + \theta)$. It is always possible to choose $\theta$ so that at some given time $t$, $\phi(t) = 0$ and consequently, $y(t) = 0$ and $a(t) = x(t)$. As a result, $a'(t) = x'(t)$. Thus the normal distribution and statistical independence of $a'$ can be attributed to the normal distribution and statistical independence of $x'$.

**Problem 4-9.** A narrow-band waveform $a(t) \cos [2\pi Ft + \phi(t)]$ is fed to a device (a detector) whose output is a function $v(a)$ of the input amplitude $a(t)$. Show that the rms bandwidth (3-42) of the output is

$$\frac{1}{2\pi}\sqrt{\frac{\mathrm{E}\,\{(dv/dt)^2\}}{\mathrm{E}\,\{v^2\}}} = \frac{1}{2\pi}\sqrt{\frac{\mathrm{E}\,\{a'^2v'^2(a)\}}{\mathrm{E}\,\{v^2(a)\}}}$$

with $v' = dv/da$ and, using (4-37), show that this is

$$\rho\sigma\sqrt{\frac{\mathrm{E}\,\{v'^2(a)\}}{\mathrm{E}\,\{v^2(a)\}}}$$

when the input is narrow-band gaussian noise. Using the results of Prob. 1-10, show that if, further, the device is a power-law detector with $v(a)$ proportional to $a^n$ for some nonnegative constant $n$, then the rms output bandwidth is $\sqrt{\frac{1}{2}n}\,\rho$, and show that the foregoing denominator must be reduced by $[\mathrm{E}\,\{v(a)\}]^2$ to obtain the bandwidth after the d-c component $\mathrm{E}\,\{v(a)\}$ is removed.[1]

**Problem 4-10.** (*a*) Describe and explain the conditional distribution $p(\phi'|a)$ of $\phi'$ for a given value of $a$.

(*b*) Determine and sketch the conditional probability density function $p(a|\phi')$. The distribution of $a$ for a given value of $\phi'$ is called a *Maxwell distribution*, and it also describes the distribution of thermal gas-molecule speeds.

**The High-Frequency Spectrum of $\phi'$.†** The infinite mean squared value of $\phi'$ results from the high probability of large instantaneous frequency deviations $\phi'$, for the tails of the distribution (4-39) fall relatively slowly compared to those of the normal distribution. These large values of $|\phi'|$, in turn, result not from high speeds of the phasor $ae^{i\phi}$ representing the noise, but rather from small values of $a$, for the tail of the Rayleigh distribution of speeds falls off rapidly.

Thus large angular velocities $|\phi'|$ of the phasor $ae^{i\phi}$ are due mainly to its passing by the origin at close range, and each such close passage results in a rapid change of $\phi$ by $\pm\pi$. If it occurs with speed $s$ at closest

[1] Phillip Bello, On the RMS Bandwidth of Nonlinearly Envelope Detected Narrow-band Gaussian Noise, *IEEE Trans. Inform. Theory*, **IT-11**:236–239 (April, 1965).

† The remainder of Chap. 4 may be omitted on the first reading.

distance $a$, it contributes to $\phi'$ a witch-of-Agnesi pulse of the form $as/(a^2 + s^2t^2)$, which is the derivative of $\phi = \arctan st/a$. The Fourier transform of this pulse is $\pi e^{-2\pi a|f|/s}$, which falls off only at very high frequencies when $a$ is small. Since small values of $a$ occur only rarely, their times of occurrence form a Poisson process, and we can apply the result of Prob. 3-13 to finding the spectral density of $\phi'(t)$ at high frequencies by multiplying the square of this Fourier transform by the rate of occurrence of impulses whose $a$ and $s$ lie within the incremental ranges $da$ and $ds$, respectively, and integrating.

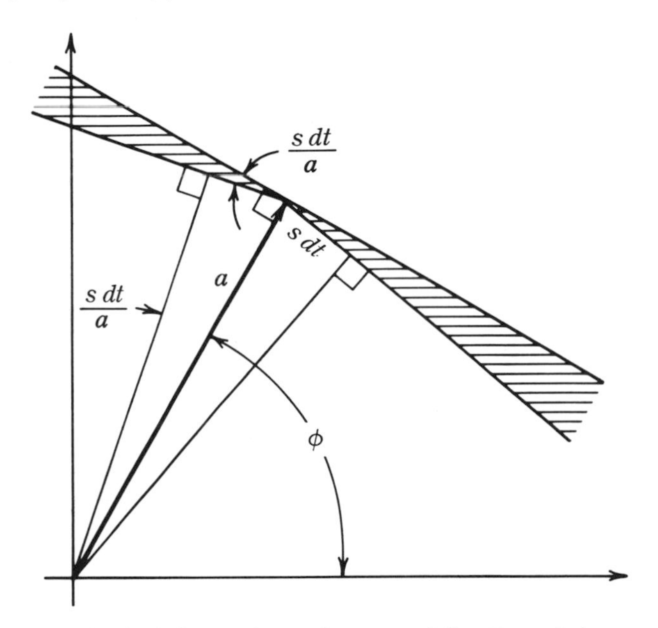


*Figure 4-3.* The shaded area shows the range of directions of phasor motion that will result in the occurrence of the middle of a pulse during a given time interval $dt$ when the phasor has magnitude $a$ and speed $s$.

To determine this rate, we evaluate the probability that the middle of a pulse will occur during a given time interval $dt$ when the noise amplitude is $a$ and the phasor speed $\sqrt{x'^2 + y'^2}$ is $s$. The middle of such a pulse occurs when the phasor is shortest, i.e., when its direction of motion is perpendicular to the phasor itself. The phasor's angular velocity is then $s/a$, and in a time $dt$ its direction changes by $s\,dt/a$. To be perpendicular to the phasor at some time during this interval, the direction of motion must, therefore, lie within an angular range $s\,dt/a$ bounded by either sense of the perpendicular to the phasor at the beginning of the interval (Fig. 4-3). With $a \ll \sigma$ we do not need to take into account the effect of changes in velocity during the pulse.

Because the direction of motion is uniformly distributed, the probability that it lies within one of these two sectors is $s\,dt/\pi a$. Dividing by $dt$, we see that the rate of occurrence of such pulses is $s/\pi a$ when the phasor has magnitude $a$ and speed $s$. Multiplying by the square of the foregoing Fourier transform and by the Rayleigh probability density functions of $a$ and $s$, which are statistically independent of each other and of the direction of motion, and integrating, we thus get, for the power spectral density of $\phi'(t)$ at high frequencies,

$$\Psi_{\phi'}(f) = \iint_0^\infty \pi^2 e^{-4\pi a|f|/s} \frac{s}{\pi a} \frac{a}{\sigma^2} e^{-a^2/2\sigma^2} \frac{s}{4\pi^2\rho^2\sigma^2} e^{-s^2/8\pi^2\rho^2\sigma^2}\, da\, ds. \tag{4-40}$$

In integrating with respect to $a$ for large $|f|$, the second exponential in (4-40) can be replaced by unity, as it is dominated by the first exponential. As a result, both integrations are easily carried out, and we get

$$\Psi_{\phi'}(f) = \frac{2\pi^2\rho^2}{|f|} \tag{4-41}$$

for large $|f|$. From (4-41) we see that $\int_{-\infty}^{\infty} \Psi_{\phi'}(f)\, df = \infty$, that is, the high-frequency tails of the spectrum of $\phi'(t)$ account for its infinite mean-squared value.

The distribution of $\phi'$ when a sinusoid of amplitude $A$ and frequency $F$ is included with the noise is rather complicated,[1] but to take account of the presence of the sinusoid in the foregoing argument we need only use (4-29) in (4-40) in place of the Rayleigh density function of $a$, thereby getting

$$\Psi_{\phi'}(f) = \frac{2\pi^2\rho^2}{|f|} \exp\left(-\frac{A^2}{2\sigma^2}\right) \tag{4-42}$$

for large $|f|$, which still gives us $\mathrm{E}\{\phi'^2\} = \infty$.

**Problem 4-11.** Show that, for narrow-band gaussian noise of rms value $\sigma$ and spectral width $\rho$, the average number of minima per second of its amplitude that are less than $a$ is $\sqrt{2\pi}\,\rho a/\sigma$ for $a \ll \sigma$, and show that the mean length of the intervals during which the amplitude is less than $a$ is, accordingly, $a/\sqrt{2\pi}\,\rho\sigma$.

**Mean Frequency Deviation of a Sinusoid Plus Noise.** In studying FM reception in the presence of noise we shall need $\mathrm{E}\{\phi'\}$ for the sum of a sinusoid $A \cos 2\pi(F + D)t$ plus narrow-band gaussian noise of rms

[1] V. P. Zhukov, Probability Density of the Phase Derivative of the Sum of a Sinusoidal Signal and Noise, *Radio Eng. Electron.* (*USSR*) (*English Transl.*) **7**:1164–1167 (1962).

value $\sigma$ whose spectrum has centroid $F$, shown in Fig. 4-1 for $t = 0$. Because $x'$ and $y'$ are independent random variables with zero mean, they contribute nothing, on the average, to $\phi'$. Thus we may regard $x$ and $y$ as fixed while the signal phasor of length $A$ rotates at the rate $2\pi D$. As long as $A$ exceeds the length $\sqrt{x^2 + y^2}$ of the noise phasor, the latter has no effect on the average rate of rotation of their sum, i.e., on the average over all phase relationships between the signal and the noise. The probability that this is the case is the integral of the Rayleigh density function (1-20) from 0 to $A$, namely, $1 - \exp(-A^2/2\sigma^2)$. (See Fig. 4-4.)

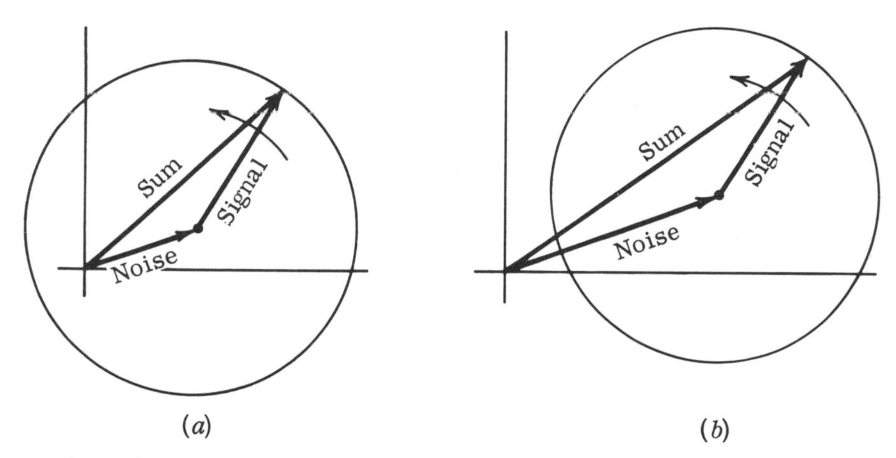


*Figure 4-4.* The rotation of the sum of a fixed noise phasor plus a rotating signal phasor when the latter is (a) longer and (b) shorter.

On the other hand, when the noise phasor is longer than the signal phasor, rotation of the latter yields no average rotation of their sum. Hence the mean value of $\phi'$ is $2\pi D$ times the probability that the noise phasor has a length less than $A$, namely,

$$\mathrm{E}\,\{\phi'\} = 2\pi D\left[1 - \exp\left(-\frac{A^2}{2\sigma^2}\right)\right]. \tag{4-43}$$

This result[1] can also be derived by noticing that the velocity $2\pi AD$ of the signal phasor in Fig. 4-1 has a component $2\pi AD \cos\phi$ normal to the total phasor $ae^{i\phi}$ which contributes $(2\pi AD/a)\cos\phi$ to $\phi'$. Using the mean value of $(\cos\phi)/a$ found in Prob. 4-7, we again get (4-43).

[1] N. M. Blachman, "The Demodulation of a Frequency-modulated Carrier and Random Noise by an FM Receiver," pp. 15–18, Cruft Laboratory, Harvard University, Tech. Rept. No. 31, March, 1948.

**Problem 4-12.** Using (4-10) and (1-12), show that, if the spectrum of a narrow-band gaussian random process $u(t) = a(t) \cos [2\pi Ft + \phi(t)]$ is symmetric[1] about the frequency $F$ and its correlation function is $\psi(\tau) \cos 2\pi F\tau$, then the conditional probability density function of $a_2 = a(t + \tau)$, given that $a(t) = a_1$, is (4-29) with $a$ replaced by $a_2$, $A$ by $\varrho a_1 = \psi(\tau)a_1/\psi(0)$, and $\sigma^2$ by $(1 - \varrho^2)\psi(0)$.

**Problem 4-13.** For the case of Prob. 4-12, show that the probability that $a_2 < ka_1$ is $\frac{1}{2} - \frac{1}{2}(1 - k^2)/\sqrt{(1 + k^2)^2 - 4\varrho^2k^2}$, by integrating the probability density function of $k = a_2/a_1$. (Note that here $a_1$ is not specified.)

**Problem 4-14.** For the case of Prob. 4-12, let $\theta$ denote the value of $\phi(t + \tau) - \phi(t)$, reduced or increased by a multiple of $2\pi$ so that $\theta$ lies between $-\pi$ and $\pi$. Show that the probability density function of $\theta$ is

$$p(\theta) = \frac{1}{2\pi} \frac{1 - \varrho^2}{1 - \varrho^2 \cos^2 \theta} \left[ 1 + \frac{\varrho \arccos(-\varrho \cos \theta)}{\sqrt{\sec^2 \theta - \varrho^2}} \right],$$

with the arccosine lying between 0 and $\pi$. The distribution of $\phi(t + \tau) - \phi(t)$ itself is an unsolved problem, to which the next section presents a partial answer.

## □ THE VARIANCE OF THE NUMBER OF ZEROS

We have already found the mean number of zeros (3-60) during an interval of length $T$ for an ergodic gaussian random process. In general, the variance of the number of zeros is given by a rather complicated formula,[2] but the narrow-band case is relatively simple,[3] provided that the spectrum is symmetric about its centroid $F$. Thus we suppose that the

[1] If the spectrum is not symmetric, we can nevertheless choose $F$ so that $z(\tau)$ is real for any given $\tau$, since changing $F$ by $\Delta F$ multiplies $z(\tau)$ by $e^{2\pi i \Delta F \tau}$. Then, by (4-23), $y(t + \tau)$ will be statistically independent of $x(t)$ as well as of $x(t + \tau)$, the latter will similarly be statistically independent of $y(t)$, and hence the results of Probs. 4-12 and 4-13 will be applicable. There is, in fact, no need here to determine an appropriate value for $F$; it suffices to substitute $|z(\tau)|$ for $\psi(\tau)$, since these two will be identical for an $F$ that makes $z(\tau)$ real, and the choice of $F$ affects neither $|z(\tau)|$ nor the amplitude $a(t)$ of the random process. See W. B. Davenport, Jr., and W. L. Root, "An Introduction to the Theory of Random Signals and Noise," pp. 162–164, McGraw-Hill Book Company, New York, 1958. In the case of Prob. 4-14, $\theta$ must, in addition, be replaced by $\theta - 2\pi \Delta F\tau$.

[2] H. Steinberg, P. M. Schultheiss, C. A. Wogrin, and F. Zweig, Short-Time Frequency Measurement of Narrow-Band Random Signals by Means of a Zero-Counting Process, *J. Appl. Phys.*, **26**:195–201 (1955); J. S. Bendat, "Random Noise Theory," pp. 392–401, John Wiley & Sons, Inc., New York, 1958.

[3] N. M. Blachman, FM Reception and the Zeros of Narrow-Band Gaussian Noise," *IEEE Trans. Inform. Theory,* **IT-10**:235–241 (July, 1964).

spectral density of the narrow-band gaussian random process $u(t)$ is

$$\Psi_u(f) = \tfrac{1}{2}\Psi(f - F) + \tfrac{1}{2}\Psi(f + F),$$

where $\Psi(f) = \Psi(-f)$ is an even function with Fourier transform $\psi(\tau)$.

By comparison with (4-8), (4-13), (4-14), and (4-23), then, we see that $u(t)$ can be expressed as

$$u(t) = x(t) \cos 2\pi Ft - y(t) \sin 2\pi Ft, \tag{4-10}$$

where $x(t)$ and $y(t)$ are independent, zero-mean, gaussian random processes, both having correlation function $\psi(\tau)$ and variance $\sigma^2 = \psi(0)$, the variance of their derivatives being (4-17) $4\pi^2\rho^2\sigma^2 = -\psi''(0)$.

**The Number of Zeros of $u(t)$.** Making use of the representation

$$u(t) = a(t) \cos [2\pi Ft + \phi(t)], \tag{4-1}$$

where

$$a(t) = \sqrt{x^2(t) + y^2(t)} \qquad \text{and} \qquad \phi(t) = \arctan \frac{y(t)}{x(t)}$$

are continuous functions of time, so that $\phi(t)$, which is determined to within a fixed additive multiple of $2\pi$, may become arbitrarily large or small, we see that $u(t)$ has a zero each time $2\pi Ft + \phi(t)$ is an odd multiple of $\frac{1}{2}\pi$. Hence, if $F$ is sufficiently large that, throughout the interval $(0,T)$,

$$\phi'(t) > -2\pi F, \tag{4-44}$$

that is, if the argument of the cosine in (4-1) does not decrease, the number of zeros $N$ of $u(t)$ in $(0,T)$ is simply the number of odd multiples of $\frac{1}{2}\pi$ lying between $\phi(0)$ and $2\pi FT + \phi(T)$, since $a(t)$, being zero only when $x(t)$ and $y(t)$ vanish simultaneously, has a zero in this interval with probability zero. Thus $N$ does not differ by more than unity from $[2\pi FT + \phi(T) - \phi(0)]/\pi$, that is,

$$\left| N - 2FT - \frac{\phi(T) - \phi(0)}{\pi} \right| < 1. \tag{4-45}$$

To simplify the calculation of $N$, we now suppose that $2FT$ is an integer. Consequently, $N$ is $2FT$ plus the number of odd multiples of $\frac{1}{2}\pi$ lying between $\phi(0)$ and $\phi(T)$, which we shall regard as positive or negative acccordingly as $\phi(T) > \phi(0)$ or $\phi(T) < \phi(0)$. Hence $N$ is $2FT$ plus the number of zeros of $x(t) = a(t) \cos \phi(t)$ in $(0,T)$ at which $\phi(t)$ is increasing, minus the number of zeros of $x(t)$ at which $\phi(t)$ is decreasing. Thus we must count each zero of $x(t)$ in the sense of

$$\operatorname{sgn} \phi' = \operatorname{sgn} (xy' - x'y)/a^2 = \operatorname{sgn} - x'y/a^2 = - \operatorname{sgn} x'y,$$

the signum being 1, 0, or $-1$ according to the sign of its argument.

Since sgn $x(t)$ changes by 2 sgn $x'$ each time $x(t)$ goes through zero and is otherwise constant, $N$ is given by the Stieltjes integral

$$N = 2FT - \tfrac{1}{2}\int_0^T \operatorname{sgn} y(t)\, d[\operatorname{sgn} x(t)]. \tag{4-46}$$

This becomes an ordinary integral if the signum is replaced by a differentiable function, and $N$ is given by its limit as that function approaches the step-function. Such a limiting process provides the justification for the first three lines of (4-47) below.

**The Variance of $N$.** Since $x(t)$ and $y(t)$ are independent, jointly ergodic random processes, the mean value of (4-46) is $2FT$. [From (3-60) we see that $E\{N\} = 2T\sqrt{F^2 + \rho^2} \cong 2FT + \rho^2 T/F$; the discrepancy is due to our assumption (4-44), which, as we shall see, results, for large $F$, in rare pairs of zeros' not being counted.] Both sgn $x(t)$ and sgn $y(t)$ have (Prob. 3-5) the correlation function $\frac{2}{\pi}\arcsin\frac{\psi(\tau)}{\sigma^2}$ with the arcsine between $-\frac{1}{2}\pi$ and $\frac{1}{2}\pi$. Thus, by (3-19), the correlation function of $\frac{d}{dt}\operatorname{sgn} x(t)$ is $-\frac{2}{\pi}\frac{d^2}{d\tau^2}\arcsin\frac{\psi(\tau)}{\sigma^2}\cdot$ The variance of (4-46), $E\{(N - 2FT)^2\}$, is therefore

$$\begin{aligned}
\operatorname{var}\{N\} &= \tfrac{1}{4}E\left\{\int_0^T dt \int_0^T ds \operatorname{sgn} y(s) \operatorname{sgn} y(t) \frac{d}{ds}\operatorname{sgn} x(s)\frac{d}{dt}\operatorname{sgn} x(t)\right\} \\
&= \tfrac{1}{4}\int_0^T dt \int_0^T ds\, E\{\operatorname{sgn} y(s)\operatorname{sgn} y(t)\}\, E\left\{\frac{d}{ds}\operatorname{sgn} x(s)\frac{d}{dt}\operatorname{sgn} x(t)\right\} \\
&= -\frac{1}{\pi^2}\int_0^T dt \int_0^T ds \arcsin\frac{\psi(t-s)}{\sigma^2}\frac{\partial^2}{\partial s^2}\arcsin\frac{\psi(t-s)}{\sigma^2} \\
&= -\frac{1}{\pi^2}\int_0^T dt\left[\arcsin\frac{\psi(t-s)}{\sigma^2}\frac{\partial}{\partial s}\arcsin\frac{\psi(t-s)}{\sigma^2}\bigg|_{s=0}^{T}\right. \\
&\qquad\left. - \int_0^T ds\left(\frac{\partial}{\partial s}\arcsin\frac{\psi(t-s)}{\sigma^2}\right)^2\right] \\
&= \frac{1}{\pi^2}\int_0^T dt\left[\arcsin\frac{\psi(t-T)}{\sigma^2}\frac{\partial}{\partial t}\arcsin\frac{\psi(t-T)}{\sigma^2}\right. \\
&\qquad\left. - \arcsin\frac{\psi(t)}{\sigma^2}\frac{d}{dt}\arcsin\frac{\psi(t)}{\sigma^2}\right] + \frac{1}{\pi^2}\iint_0^T \frac{\psi'^2(t-s)\,ds\,dt}{\sigma^4 - \psi^2(t-s)} \\
&= \frac{1}{4} - \frac{1}{\pi^2}\arcsin^2\frac{\psi(T)}{\sigma^2} + \frac{2}{\pi^2}\int_0^T (T-\tau)\frac{\psi'^2(\tau)\,d\tau}{\sigma^4 - \psi^2(\tau)}
\end{aligned} \tag{4-47}$$

The first three lines of (4-47) must be justified by the limiting process mentioned above, because $(\partial/\partial s)\arcsin\psi(t-s)/\sigma^2 = \mp 2\pi\rho$ for $s = t \pm 0$.

**The Case of Nonintegral $2FT$.** In general, it is difficult to determine var $\{N\}$ when $2FT$ is not an integer. However, with $F$ large, reducing $T$ to the next smaller value—$[2FT]/2F$—that makes $2FT$ an integer reduces $N$ by either 0 or 1. The variance of this change in $N$ cannot exceed $\frac{1}{4}$, and its covariance with the remainder of $N$ must lie between $-\frac{1}{4}$ and 0. This is so because the change in $N$ depends only on $F$, $T$, and $\phi(T)$, while $N$ itself depends only on $FT$, $\phi(0)$, and $\phi(T)$ [and $\phi(T) - \phi(0)$ is distributed symmetrically about zero because of the symmetry of arctan $y(t)/x(t)$]. Hence, when $2FT$ is not an integer, var $\{N\}$ does not differ by more than $\frac{1}{4}$ from (4-47), which is altered negligibly when $T$ is replaced by $[2FT]/2F$ if $F$ is large.

When $\psi(\tau)$ is a constant and $u(t)$ is, as (2-3) shows, a sinusoid of frequency $F$, the exact value of var $\{N\}$ is easily shown to be

$$\text{var}\ \{N\} = (2FT - [2FT])(1 - 2FT + [2FT]), \tag{4-48}$$

which oscillates like a bouncing ball between 0 and $\frac{1}{4}$ with period $1/2F$. Here $[2FT]$ is the largest integer not exceeding $2FT$. For any $\psi(\tau)$, as long as $T \ll 1/\rho$, $u(t)$ is indistinguishable from a sinusoid, and (4-48) is applicable. For $T \gg 1/\rho$ [because $u(t)$ near $t = T$ generally becomes statistically independent of $u(t)$ near $t = 0$], var $\{N\}$ no longer oscillates in this manner but is very close to (4-47).

**Problem 4-15.** Prove (4-48) by noticing that, with $\psi(\tau)$ constant, $\phi(t)$ is a uniformly distributed constant and that $N$ is $[2FT]$ or $[2FT] + 1$ accordingly as an odd multiple of $\frac{1}{2}\pi$ does not or does fall between $\pi[2FT] + \phi(T)$ and $2\pi FT + \phi(T)$.

**Problem 4-16.** If $T$ increases by $P/2F$ from a value for which $2FT$ is an integer, show that, for $0 \le P \le 1$, the resulting increase $\Delta N$ in $N$ has variance var $\{\Delta N\} = P - P^2$ (see Prob. 4-15).

**Problem 4-17.** Recalling that, with $2FT$ an integer, $N$ is $2FT$ plus the number $M$ of odd multiples of $\frac{1}{2}\pi$ falling between $\phi(0)$ and $\phi(T)$, with E $\{M\} = 0$, and averaging over the uniform distribution of $\phi(T)$, show that the covariance of $N$ and $\Delta N$ (see Prob. 4-16) is E $\{M\ \Delta N\}$ and that it must lie between $-\frac{1}{2}P$ and 0 for $0 \le P \le \frac{1}{2}$ or between $\frac{1}{2}(P - 1)$ and 0 for $\frac{1}{2} \le P \le 1$, the extremes being the cases where $|\phi(T) - \phi(0)|$ is always $\frac{1}{2}\pi$ or is always 0, respectively. Thus, var $\{N + \Delta N\}$ = var $\{N\}$ + 2 cov $\{N, \Delta N\}$ + var $\{\Delta N\}$ lies between var $\{N\} + P - P^2$ and var $\{N\} - P^2$ or var $\{N\} - (1 - P)^2$, that is, between var $\{N\} + \frac{1}{4}$ and var $\{N\} - \frac{1}{4}$.

**The Effect of $\phi' < -2\pi F$.** The difference, approximately $\rho^2 T/F$, between E $\{N\} = 2T\sqrt{F^2 + \rho^2}$ and the mean of (4-46), $2FT$, results, as we have noted, from the fact that (4-44) may not hold throughout the interval $(0,T)$. The probability of its violation is not extremely low, for the tails of the distribution (4-39) of $\phi'$ fall off relatively slowly. Like the infinite value of E $\{\phi'^2\}$, the violations of (4-44) are attributable to the phasor $x(t) + iy(t)$ passing by the origin at close range and contributing to $\phi'(t)$ a witch-of-Agnesi pulse of area $\pm\pi$.

The interval during which (4-44) is violated in connection with such an event encompasses a change of less than $\pi$ in the value of $\phi(t)$. It can therefore cause the argument of the cosine in (4-1) to descend through no more than one odd multiple of $\frac{1}{2}\pi$, thereby increasing $N$ by two—one for the descent and one for the ascent through the same value that necessarily follows immediately.

The difference $\rho^2T/F$ thus indicates that two zeros can be expected to be added to $u(t)$ in this way on $\rho^2T/2F$ such occasions. [This is one-quarter of the expected number of peaks of $\phi'(t)$ violating (4-44); see Prob. 4-18 below.] Since such events are rare for large $F$, they form a Poisson process, and the variance of their number (see Prob. 3-10) equals its mean, $\rho^2T/2F$. Thus, multiplying by the square of the amount each such event contributes to $N$, we get a total resulting contribution $2\rho^2T/F$, which should be added to (4-47) as a first-order correction for the finiteness of $F$. This correction term will be small compared to (4-47) whenever $\rho \ll F$.

**Problem 4-18.** By a method similar to that used in (4-40) and in Prob. 4-11, show that, for narrow-band gaussian noise of spectral width $\rho$, the average number of times per second that its instantaneous frequency deviation $\phi'(t)$ exceeds $2\pi D$ is $2\rho^2/D$ for $D \gg \rho$.

**Approximation for Large $T$.** When $\rho T$ is small, (4-47) becomes approximately $2\rho T$, the expected number of zeros of $x(t)$. For large $T$, a good approximation to var $\{N\}$ is $T$ times

$$V(0) = \lim_{T\to\infty} \frac{\text{var}\,\{N\}}{T} = \frac{2}{\pi^2}\int_0^\infty \frac{\psi'^2\,d\tau}{\sigma^4 - \psi^2}, \tag{4-49}$$

which is shown in the table on the facing page for several power spectral densities.

**Narrow-Band Gaussian Noise Plus a Sinusoid.** Although it is possible to find var $\{N\}$ for the sum of a sinusoid of frequency $F$ plus narrow-band gaussian noise with spectrum symmetric about this frequency, we shall not need the exact result,[1] which is somewhat com-

[1] Blachman, *ibid.*

plicated. Instead, using a technique due to Robert Price, we shall merely determine $\lim_{T\to\infty} T^{-1} \operatorname{var}\{N\}$, which is independent of the phase of the sinusoid and takes a simpler form.

Thus, increasing $\psi(\tau)$ by the constant $1/s$, we find that (4-49) becomes

$$\frac{2}{\pi^2}\int_0^\infty \frac{\psi'^2\,d\tau}{\left(\sigma^2+\frac{1}{s}\right)^2-\left(\psi+\frac{1}{s}\right)^2} = \frac{2}{\pi^2}\int_0^\infty \frac{\psi'^2\,d\tau}{\sigma^4-\psi^2+\frac{2}{s}(\sigma^2-\psi)}. \tag{4-50}$$

When multiplied by $T$, (4-50) gives the variance of the number of zeros for large $T$ when the spectrum of $u(t)$ consists of a line of strength $1/s$ on frequency $F$ plus the spectrum corresponding to $\psi(\tau)$, which is symmetric about $F$. The spectral line manifests itself by the presence, in the random process, of a sinusoidal component $A\cos(2\pi Ft+\alpha)$ whose constant amplitude $A$ is Rayleigh-distributed with probability density function $sA\exp(-\frac{1}{2}sA^2)$ and whose constant phase $\alpha$ is uniformly distributed. [See (2-1).]

Letting

$$V(A) = \lim_{T\to\infty}\frac{\operatorname{var} N}{T}, \tag{4-51}$$

where $N$ is the number of zeros in $(0,T)$ of the sum of a sinusoid of amplitude $A$ and frequency $F$ plus narrow-band gaussian noise with

| $\Psi(f)$ | $\psi(\tau)$ | $V(0)$ |
|---|---|---|
| $\frac{\sigma^2}{\rho}\operatorname{sech}\frac{\pi f}{\rho}$ | $\sigma^2\operatorname{sech}\pi\rho\tau$ | $\frac{2\rho}{\pi} = 0.6366\rho$ |
| $\frac{\sigma^2}{\rho}K_0\left(\frac{\|f\|}{\rho}\right)$ | $\frac{\sigma^2}{\sqrt{1+4\pi^2\rho^2\tau^2}}$ | $\rho$ |
| $\frac{\sigma^2}{\sqrt{2}\,\rho}\exp\left(-\frac{\sqrt{2}\,\|f\|}{\rho}\right)$ | $\frac{\sigma^2}{1+2\pi^2\rho^2\tau^2}$ | $(4-\sqrt{8})\rho = 1.1716\rho$ |
| $\frac{\sigma^2}{\sqrt{2\pi}\,\rho}\exp\left(-\frac{f^2}{2\rho^2}\right)$ | $\sigma^2\exp(-2\pi^2\rho^2\tau^2)$ | $\frac{\zeta(\frac{3}{2})}{\sqrt{\pi}}\rho = 1.475\rho$ |
| $\frac{\sigma^2}{B}$ for $\|f\| < \frac{1}{2}B = \sqrt{3}\,\rho$,<br>0 for $\|f\| > \frac{1}{2}B$ | $\sigma^2\operatorname{sinc} B\tau$ | $0.612B = 2.12\rho$ |
| $\frac{1}{2}\sigma^2\delta(f-\rho)+\frac{1}{2}\sigma^2\delta(f+\rho)$ | $\sigma^2\cos 2\pi\rho\tau$ | $\infty$ |

correlation function $\psi(\tau)\cos 2\pi F\tau$, we see, by a slight generalization of (4-46), that $\mathrm{E}\{N\} = 2FT$ for any $A$ when $F$ is large. Hence (4-50) must be the average of $V(A)$ over the Rayleigh distribution of $A$, namely,

$$s\int_0^\infty AV(A)e^{-\frac{1}{2}sA^2}\,dA = s\int_0^\infty V(\sqrt{2p})e^{-sp}\,dp, \tag{4-52}$$

where $p = \frac{1}{2}A^2$ is the mean-squared value of the sinusoid. Recognizing (4-52) as a Laplace transform, we conclude that $V(\sqrt{2p})$ is the inverse Laplace transform of $1/s$ times (4-50), namely,

$$\frac{2}{\pi^2}\int_0^\infty \frac{\psi'^2\,d\tau}{(\sigma^4 - \psi^2)s + 2(\sigma^2 - \psi)}.$$

Since the inverse Laplace transform of $1/(s + a)$ is $e^{-ap}$, it follows that

$$V(\sqrt{2p}) = \frac{2}{\pi^2}\int_0^\infty \frac{\psi'^2}{\sigma^4 - \psi^2}\exp\left(-\frac{2p}{\sigma^2 + \psi}\right)d\tau$$

or

$$V(A) = \frac{2}{\pi^2}\int_0^\infty \frac{\psi'^2}{\sigma^4 - \psi^2}\exp\left(-\frac{A^2}{\sigma^2 + \psi}\right)d\tau. \tag{4-53}$$

The variance of $N$ is approximately $TV(A)$ and, in fact, differs by no more than unity from (4-53) with a factor $(T - \tau)$ included in the integrand and with the upper limit changed to $T$.

**Spectral Density of $\phi'(t)$.** Applying (3-10) to $\phi'(t)$, we see that its spectral density for $f = 0$, $\Psi_{\phi'}(0)$, is the limit of

$$\begin{aligned} T^{-1}\,\mathrm{E}\left\{\left[\int_{-T/2}^{T/2}\phi'(t)\,dt\right]^2\right\} &= T^{-1}\,\mathrm{E}\left\{\left[\int_0^T \phi'(t)\,dt\right]^2\right\} \\ &= T^{-1}\,\mathrm{E}\,\{[\phi(T) - \phi(0)]^2\}, \end{aligned}$$

since $\int_{-T/2}^{T/2}\phi'(t)\,dt$ is the Fourier transform of a sample of $\phi'(t)$ of duration $T$, evaluated at $f = 0$. Thus, inasmuch as

$$\mathrm{E}\,\{\phi(T) - \phi(0)\} = 0,$$

we have

$$\Psi_{\phi'}(0) = \lim_{T\to\infty}\frac{\operatorname{var}\{\phi(T) - \phi(0)\}}{T}.$$

From (4-45) we see that, for large $T$, the variance of $\phi(T) - \phi(0)$ is approximately $\pi^2$ times the variance of $N$. Hence, passing to the limit with the help of (4-51), we find that

$$\Psi_{\phi'}(0) = \pi^2 V(A). \tag{4-54}$$

[See (4-42).]

**The Case of Large $A$.** When $A \gg \sigma$, nearly all of (4-53) comes from small values of $\tau$. We can therefore approximate it by using the first two terms, $\sigma^2 - 2\pi^2\rho^2\sigma^2\tau^2$, of the power series for $\psi(\tau)$, from which we get

$$V(A) = \frac{8\rho\sigma}{\sqrt{2\pi}\,A} \exp\left(-\frac{A^2}{2\sigma^2}\right). \tag{4-55}$$

(In the case of a periodic correlation function, $V(A)$ is infinite, since (4-53) then contains (4-55) many times over.) We can also derive (4-55) directly by noticing that for large $T$ the phase $\alpha$ of the sinusoid matters very little and can be set equal to zero, giving (Fig. 4-1)

$$\begin{aligned} u(t) &= [A + x(t)] \cos 2\pi Ft - y(t) \sin 2\pi Ft \\ &= a(t) \cos [2\pi Ft + \phi(t)] \end{aligned} \tag{4-56}$$

with
$$\phi(t) = \arctan \frac{y(t)}{A + x(t)}. \tag{4-57}$$

When $A \gg \sigma$, (4-57) is nearly always close to a multiple of $2\pi$, jumping on rare occasions from one multiple to the next higher or lower. The variance of $N$ is then due almost entirely to these jumps of $\pm 2\pi$. Each such jump adds $\pm 2$ to $N$, and since upward and downward jumps are, by symmetry, equally likely, each contributes $(\pm 2)^2 = 4$ to the variance of $N$.

Such a jump can occur only if $y$ goes through zero while $x < -A$. Thus, multiplying $2\rho$, the expected number of zeros of $y$ per unit time, by erf $(-A/\sigma)$, the probability that $x < -A$ given by (1-5), and by $(\pm 2)^2 = 4$, we get the approximation

$$V(A) = 8\rho \operatorname{erf}(-A/\sigma), \tag{4-58}$$

which becomes (4-55) upon substitution of the first term of the asymptotic series (1-6) for the error function.

Since $(d/dt)$ E $\{xx'\} = 0$, the covariance of $x(t)$ and $x''(t)$ is

$$\mathrm{E}\,\{xx''\} = -\mathrm{E}\,\{x'^2\} = -4\pi^2\rho^2\sigma^2.$$

Hence, by (1-14), if $x(t)$ is very negative, $x''(t)$ is likely to be very large. Therefore, when $A$ is large and $x(t)$ passes downward through the value $-A$, it quickly turns and passes upward again through this value. In the rare case that in the interim $y(t)$ has passed through zero, $\phi(t)$ will have changed very rapidly by $\pm 2\pi$. This will not occur, however, if $A$ is not large. Instead, as we saw earlier, $\phi(t)$ will relatively frequently change quickly by $\pm\pi$ as the phasor $x + iy$ passes close to the origin.

**Problem 4-19.** For the sum $a(t) \cos [2\pi(F + D)t + \phi(t)]$ of a strong sinusoid $A \cos 2\pi(F + D)t$ plus narrow-band gaussian noise of rms value $\sigma$ and spectrum of width $\rho$ symmetric about frequency $F$, show that the expected number of upward jumps of $\phi(t)$ per unit time is[1]

$$\rho\sqrt{1+\left(\frac{D}{\rho}\right)^2}\,\mathrm{erf}\left(-\frac{A}{\sigma}\sqrt{1+\left(\frac{D}{\rho}\right)^2}\right) - De^{-A^2/2\sigma^2}\,\mathrm{erf}\left(-\frac{AD}{\rho\sigma}\right)$$

by noticing that the probability of an upward jump between $t = 0$ and $t = dt$ is the probability that, at time $t = 0$, $x$ is less than $-A$, that $y'$ is less than $2\pi Dx$, and that $y$ lies between 0 and $(2\pi Dx - y')\,dt$, since changing from reference frequency $F$ to $F + D$ adds an angular velocity $-2\pi D$ to our noise phasor $x + iy$, thus contributing a downward velocity $2\pi Dx$ to the signal-plus-noise phasor. Proceeding as in the derivation of (3-60), integrate (4-36) over all $x'$ and over all $y, y'$, and $x$ satisfying these conditions, and divide by $dt$. The $x'$ and $y$ integrations are trivial. The remaining two integrations are facilitated by substituting $z + 2\pi Dx$ for the $y'$ in the exponent and expressing the factor $(y' - 2\pi Dx)$ as $z = (1 + D^2/\rho^2)(z + 2\pi Dx) - 2\pi D(x + D^2x/\rho^2 + Dz/2\pi\rho^2)$. The first of the two resulting terms is easily integrated with respect to $z$ and then $x$, and the second term is easily integrated with respect to $x$ and then $z$.

**Problem 4-20.** From the result of Prob. 4-19 show that the expected number of downward jumps of $\phi(t)$ per unit time is

$$\rho\sqrt{1+\left(\frac{D}{\rho}\right)^2}\,\mathrm{erf}\left(-\frac{A}{\sigma}\sqrt{1+\left(\frac{D}{\rho}\right)^2}\right) + De^{-A^2/2\sigma^2}\left[1 - \mathrm{erf}\left(-\frac{AD}{\rho\sigma}\right)\right]$$

and that hence

$$\Psi_{\phi'}(0+) = 8\pi^2\rho\sqrt{1+\left(\frac{D}{\rho}\right)^2}\,\mathrm{erf}\left(-\frac{A}{\sigma}\sqrt{1+\left(\frac{D}{\rho}\right)^2}\right) + 4\pi^2De^{-A^2/2\sigma^2}\left[1 - 2\,\mathrm{erf}\left(-\frac{AD}{\rho\sigma}\right)\right].$$

Note that the difference between the expected number of upward and downward jumps per unit time is the average number of revolutions per second of the signal-plus-noise phasor about the signal phasor, $-D\exp(-A^2/2\sigma^2)$, and that the average frequency deviation of the sum of the signal plus the noise is hence not $D$ but $[1 - \exp(-A^2/2\sigma^2)]D$ [see (4-43)]. This $\Psi_{\phi'}(0+)$ corresponds to the first term of the spectrum of Prob. 3-13 with $\sigma^2 + \mu^2 = \mathrm{E}\{a_j^2\} = 4\pi^2$ and $v(t) = \delta(t)$; the net average number of jumps per unit time corresponds to the second term, which represents $\mathrm{E}\{\phi'\}$.

[1] S. O. Rice, Noise in FM Receivers in M. Rosenblatt (ed.), "Time Series Analysis" chap. 25, pp. 395–422, John Wiley & Sons, Inc., New York, 1963.

# PART II

# Demodulation, Detection, and Other Nonlinearities

# chapter 5 NONLINEAR DEVICES

If a narrow-band waveform

$$u(t) = a(t) \cos [2\pi Ft + \phi(t)]$$

representing a signal plus noise, for example, is fed to a nonlinear device whose output $v$ is a function $v(u)$ of its input $u$ at the same instant, the output will be $v(a \cos \theta)$, where $\theta = 2\pi Ft + \phi$. The nonlinear device may be a detector (i.e., an amplitude demodulator), a harmonic generator, a limiter, or simply the imperfect realization of a linear amplifier.

Since the output $v(a \cos \theta)$ is an even, periodic function of $\theta$, it is represented exactly, for all $\theta$, by the Fourier series[1]

$$v(a \cos \theta) = \tfrac{1}{2}v_0 + v_1 \cos \theta + v_2 \cos 2\theta + \cdots,$$

where

$$v_m(a) = \frac{1}{\pi} \int_0^{2\pi} v(a \cos \theta) \cos m\theta \, d\theta, \tag{5-1}$$

these Fourier coefficients being functions of the amplitude $a(t)$ of the input waveform. Substituting $\theta = 2\pi Ft + \phi$, we can thus express the output as

$$v = \tfrac{1}{2}v_0(a) + v_1(a) \cos (2\pi Ft + \phi) + v_2(a) \cos (4\pi Ft + 2\phi) + \cdots. \tag{5-2}$$

Here we see that the first term, $\frac{1}{2}v_0(a)$, represents the low-frequency component of the output—the audio or video output—which is the only important component in the case of detection. The next term,

$$v_1(a) \cos (2\pi Ft + \phi),$$

[1] N. M. Blachman, Band-Pass Nonlinearities, *IEEE Trans. Inform. Theory*, **IT-10**:162–164 (1964).

represents the input-frequency component of the output and is the only important term in the case of a nonlinear amplifier or limiter. Notice that it reproduces the input phase $\phi(t)$ exactly but distorts the input amplitude.

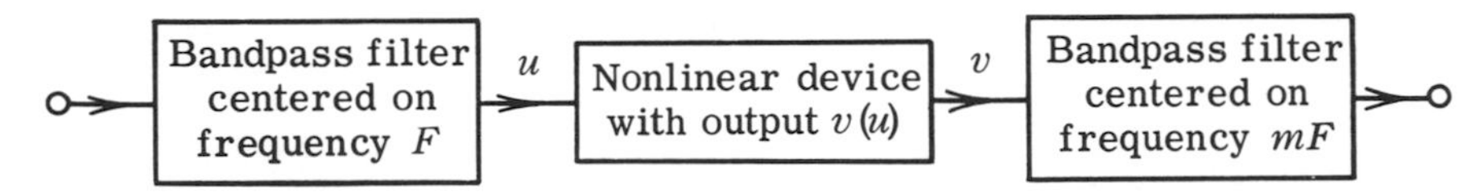


*Figure 5-1.* Nonlinear device with narrow-band input followed by a filter passing the $m$th-harmonic zone.

The remaining terms of (5-2) represent harmonics of the input. If the nonlinear device is followed by a filter (Fig. 5-1) that passes all frequencies in the neighborhood of the $m$th harmonic of its input, only the term $v_m(a) \cos(2m\pi Ft + m\phi)$ will appear in the filter's output.

**Problem 5-1.** Show that the *odd part* $\frac{1}{2}v(u) - \frac{1}{2}v(-u)$ of any nonlinearity $v(u)$ never yields any low-frequency output $\frac{1}{2}v_0(a)$, and show that, to obtain a low-frequency output equal to the input amplitude $a(t)$, it suffices to use a "linear" detector, i.e., a nonlinear device with $v(u) = \frac{1}{2}\pi|u|$ (a full-wave linear rectifier) or with $v(u) = \frac{1}{2}\pi(|u| + u)$ (a half-wave linear rectifier).

**Problem 5-2.** Show that, to obtain a low-frequency detector output equal to $\ln I_0(a)$, one can use a detector with $v(u) = \frac{1}{2}u^2 - \frac{1}{18}u^4 + \frac{1}{180}u^6 - \frac{11}{13,440}u^8 + \frac{19}{75,600}u^{10} - \cdots$. (HINT: Obtain the power series for $\ln I_0(a)$ by dividing the power series for $I_1(a)$ by that for $I_0(a)$ and integrating.) This series converges only for $|u| < 2.4048$; by using (4-28), show that, for large $|u|$, $v(u) = \frac{1}{2}\pi|u| - \frac{1}{2}\ln|u| + O(1)$. Such a nonlinearity will be needed in connection with Fig. 7-3.

**Signal and Noise Output.** If the input $u(t)$ consists of a signal $A(t)\cos[2\pi Ft + \alpha(t)]$ and noise which are jointly ergodic, we can define the *gross output signal* as $\mathrm{E}\{v|A,\alpha\}$, the conditional mean of the output $v(t)$ for given values of the input-signal amplitude $A(t)$ and phase $\alpha(t)$ at the same instant. This quantity depends on $t$ through $A$ and $\alpha$, being the average of $v(t)$ over the ensemble statistics of the noise. This definition is justified by the fact that, if the input signal is periodic or its amplitude varies periodically, we associate with it the output components having the same period, and these can be separated from the remainder of the output by means of a "comb filter," which passes only the harmonics (including direct current and the fundamental) of the reciprocal of the period. Such a filter, integrating [see (3-3)] over many repetitions of the input signal because of its small bandwidth, averages out the output fluctuations due to the input noise, thereby yielding E $\{v|A,\alpha\}$ regardless of the period of the input signal.

This gross output signal includes direct current, low frequencies, and input-signal harmonics, but by taking the appropriate term of (5-2) instead of the total output $v$, we obtain the output signal of interest in any particular application, for example, E $\{v_m(a) \cos(2m\pi Ft + m\phi)|A,\alpha\}$. Notice that this $m$th-harmonic output signal is a definite function of the amplitude $A$ and phase $\alpha$ of the input signal. Thus it is the output that would be obtained if the same signal, without the noise, were fed to a different nonlinear device. (See Prob. 5-4.)

The fluctuations of the output (5-2) [or any term of (5-2)] about its conditional mean may be called the *output noise*, $v$ − E $\{v|A,\alpha\}$. It is orthogonal to the output signal in the sense that the expectation of their product (averaged over both signal and noise statistics) is zero, i.e., they are uncorrelated. Hence the output noise power[1] E $\{[v - \mathrm{E}\,\{v|A,\alpha\}]^2\}$ is equal to the difference between the total output power E $\{v^2\}$ and the output signal power E $\{[\mathrm{E}\,\{v|A,\alpha\}]^2\}$, and the output signal-to-noise ratio is

$$R = \frac{\mathrm{E}\,\{[\mathrm{E}\,\{v|A,\alpha\}]^2\}}{\mathrm{E}\,\{v^2\} - \mathrm{E}\,\{[\mathrm{E}\,\{v|A,\alpha\}]^2\}}. \tag{5-3}$$

**Problem 5-3.** Determine the gross output signal E $\{v|A,\alpha\}$ of an ideal limiter with output $v = \operatorname{sgn} u$ whose input $u$ is a signal plus zero-mean gaussian noise of rms value $\sigma$.

**Problem 5-4.** Show that, if a signal plus zero-mean gaussian noise of power $\sigma^2$ is passed through an error-function limiter (see Prob. 3-6) with "variance" $\Sigma^2$, the gross output signal (or any of its harmonic components) has exactly the same form as if there were no noise and the limiter's "variance" were $\Sigma^2 + \sigma^2$—or the limiter were ideal ($\Sigma = 0$) and the noise power were[2] $\sigma^2 + \Sigma^2$.

**Problem 5-5.** For an exponential device with output $v = e^u$ whose input $u$ is a signal $A \cos 2\pi Ft$ plus narrow-band gaussian noise of power $\sigma^2$, use (1-27) to show that the gross output signal is $e^{A \cos 2\pi Ft + \sigma^2/2}$, and show that it can be resolved into a d-c component of power $I_0^2(A)e^{\sigma^2}$ plus a component of every harmonic frequency $mF$ ($m = 1, 2, \ldots$) of power $2I_m^2(A)e^{\sigma^2}$. Show that the total noise-output power is $I_0(2A)(e^{2\sigma^2} - e^{\sigma^2})$, of which E $\{I_0^2(a)\} - I_0^2(A)e^{\sigma^2}$

[1] The first E here averages over the statistics of both signal and noise; in the expression for the output signal power it averages over the statistics of the signal's amplitude and phase.

[2] N. M. Blachman, The Effect of a Limiter upon Signals in the Presence of Noise, *IRE Trans. Inform. Theory*, **IT-6**; 52 (1960). See also Robert Price, A Useful Theorem for Nonlinear Devices Having Gaussian Inputs, *IRE Trans. Inform. Theory*, **IT-4**(2): 69–72 (June, 1958).

represents the low-frequency output noise and $2\mathrm{E}\,\{I_m^2(a)\} - 2I_m^2(A)e^{\sigma^2}$ represents the $m$th-harmonic output noise, where E denotes the average over the distribution (4-29), which evidently requires numerical integration in the present case.

**The Low-Frequency Output Signal, Distortion, and Direct Current.** For $v$ in (5-3) we can substitute any appropriate term of (5-2) except the first, $\frac{1}{2}v_0(a)$, for it includes the d-c output component $\mathrm{E}\,\{v\} = \mathrm{E}\,\{\frac{1}{2}v_0(a)\}$, which, being constant, conveys no information and is hence generally not regarded as a part of the output signal. In fact, the output usually contains a d-c component even when there is no input signal.

Thus, subtracting the d-c output, we find that the low-frequency output signal is, say,

$$h = \mathrm{E}\,\{\tfrac{1}{2}v_0(a)|A\} - \mathrm{E}\,\{\tfrac{1}{2}v_0(a)\}. \tag{5-4}$$

Since $\mathrm{E}\,\{h\} = 0$, the low-frequency output signal is orthogonal to the d-c output, and we find its average power by subtracting the d-c output power $[\mathrm{E}\,\{\frac{1}{2}v_0(a)\}]^2$ from the gross low-frequency output signal power $\mathrm{E}\,\{[\mathrm{E}\,\{\frac{1}{2}v_0(a)|A\}]^2\}$. The low-frequency output signal-to-noise ratio is therefore

$$R = \frac{\mathrm{E}\,\{[\mathrm{E}\,\{v_0(a)|A\}]^2\} - [\mathrm{E}\,\{v_0(a)\}]^2}{\mathrm{E}\,\{v_0^2(a)\} - \mathrm{E}\,\{[\mathrm{E}\,\{v_0(a)|A\}]^2\}}. \tag{5-5}$$

The low-frequency output signal (5-4) of a radio receiver is generally intended to reproduce some particular waveform $g(t)$, such as a microphone displacement or picture-element brightness, which is available at the transmitter and which is made to modulate the transmitted signal $A(t)\cos[2\pi Ft + \alpha(t)]$ in some manner. Hence we separate $h(t)$, (5-4), into two components, one reproducing $g(t)$ and the other orthogonal to it. If the first component is $cg(t)$, the second must be $h(t) - cg(t)$, and we thus have $\mathrm{E}\,\{g(t)[h(t) - cg(t)]\} = 0$. Hence $c = \mathrm{E}\,\{g(t)h(t)\}/\mathrm{E}\,\{g^2(t)\}$, and the undistorted output signal is

$$\frac{\mathrm{E}\,\{g(t)h(t)\}}{\mathrm{E}\,\{g^2(t)\}}\,g(t). \tag{5-6}$$

The remainder of the gross low-frequency output signal, the difference between $h(t)$ and (5-6), represents distortion (and intermodulation). If $g(t)$ is the sum of sinusoids of incommensurable frequencies, this remainder will consist of their second and higher harmonics and of sum and difference frequencies—all orthogonal to $g(t)$.

In this analysis of $h(t)$ we have tacitly supposed that any delay in the transmission of $g(t)$ has been taken into account; in the case of a multipath transmission medium, $g(t)$ suffers more than one delay, and a more complicated analysis is needed, taking into account the use to which the received signal is put. Notice that our assumption has enabled us to resolve the low-frequency output $\frac{1}{2}v_0$ into four mutually orthogonal components: undistorted signal, distortion, noise, and direct current. All of these are, of course, orthogonal to the fundamental and harmonic terms of (5-2).

**Problem 5-6.** Show that the average power of the distortion, $\mathrm{E}\{[h(t) - cg(t)]^2\}$, is minimized by the foregoing value of $c$.

**Problem 5-7.** For the case where $\frac{1}{2}v_0(a) = \frac{1}{2}a^2$ and the input $u(t)$ is the sum of the amplitude-modulated signal $A[1 + mg(t)] \cos (2\pi Ft + \theta)$ plus narrow-band gaussian noise of power $\sigma^2$, find the undistorted output-signal power and distortion power ($a$) when $g(t)$ is a sinusoid of unit rms value and ($b$) when $g(t)$ is a gaussian random process of zero mean and unit variance. Show that the ratio of the undistorted output-signal power to the low-frequency output-noise power is $m^2A^4/(\sigma^4 + \sigma^2A^2 + m^2\sigma^2A^2)$ in either case, being $8/(8 + m^2)$ and $2/(2 + m^2)$ times (5-5), respectively.

## ☐ THE POWER - LAW RECTIFIER

To obtain more specific results, we shall now apply the foregoing approach to the case of a half-wave power-law rectifier[1] which, with input $u$, gives output (Fig. 5-2)

$$v = \begin{cases} u^n & \text{for } u > 0, \\ 0 & \text{for } u \leq 0, \end{cases} \tag{5-7}$$

where $n$ is any nonnegative constant—not necessarily an integer. For $n = 1$ and $n = 2$, (5-7) represents a "linear" and a square-law rectifier, respectively, and for $n = 0$ it represents an ideal limiter with $v = 1$ for all positive values of $u$ and $v = 0$ for negative values, so that all input amplitude fluctuations are removed. Values of $n$ between 0 and 1 result in imperfect limiting, as the output amplitude then fluctuates less than the input.

We shall first obtain general formulas for the output signal, noise, and so on, valid for any power-law exponent $n$ and applicable to any output harmonic zone $m$, since this generality costs very little extra difficulty apart from the appearance of the unfamiliar confluent hypergeometric functions. Afterward we shall substitute particular values for $m$ and $n$

[1] N. M. Blachman, The Output Signal-to-Noise Ratio of a Power-Law Device, *J. Appl. Phys.*, **24**:783–785 (1953). See also W. B. Davenport, Signal-to-Noise Ratios in Band-Pass Limiters, *J. Appl. Phys.*, **24**:702–727 (1953).

to see what specific forms these quantities take, for example, in the cases of the linear and square-law detection of amplitude-modulated signals in the presence of gaussian noise. We shall also see in general and in particular how the formula for the output signal-to-noise ratio simplifies when the input signal is very weak, and when it is very strong, compared to the input noise. Later we shall study the finer details of the output spectrum, since it is often not sufficient to know, for example, the total low-frequency output noise power.

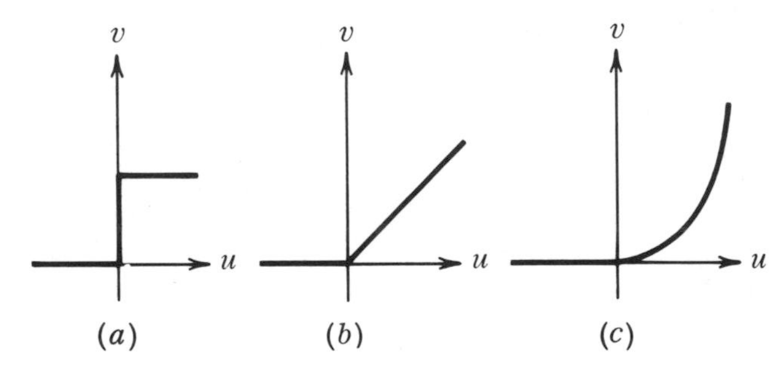


*Figure 5-2.* Output $v$ versus input $u$ for $(a)$ a limiter $(n = 0)$, $(b)$ a half-wave linear rectifier $(n = 1)$, and $(c)$ a half-wave square-law rectifier $(n = 2)$.

Substituting (5-7) into (5-1), we find[1] that

$$v_m(a) = b_m(n)a^n, \tag{5-8}$$

where

$$b_m(n) = 2^{-n}\binom{n}{\frac{1}{2}n - \frac{1}{2}m} = \frac{2^{-n}n!}{(\frac{1}{2}n - \frac{1}{2}m)!(\frac{1}{2}n + \frac{1}{2}m)!}. \tag{5-9}$$

Here again $s!$ denotes $\Gamma(s + 1)$, (1-7), when $s$ is not an integer; the parenthesis on the first line of (5-9) represents a binomial coefficient. When $n$ is an integer, these coefficients vanish for

$$m = n + 2, n + 4, n + 6, \ldots,$$

and the corresponding harmonic terms do not appear in the output (5-2).

**Half-Wave and Full-Wave Power-Law Devices.** Corresponding to (5-7) is the full-wave rectifier (Fig. 5-3) with $v = |u|^n$ for all $u$. Its substitution into (5-1) shows that all even harmonics (including low frequencies) are doubled in magnitude, while all odd harmonics disappear.

[1] Wolfgang Gröbner and Nikolaus Hofreiter, "Integraltafel, 2. Teil, Bestimmte Integrale," 2d ed., formula 335.19, Springer-Verlag OHG, Vienna, 1958. Or Bateman Manuscript Project, "Tables of Integral Transforms," vol. 1, p. 22, No. (27), McGraw-Hill Book Company, New York, 1954.

Similarly, a full-wave $n$th-power device of odd symmetry with $v = u^n$ for $u > 0$ and $v = -(-u)^n$ for $u < 0$ results in the doubling of all *odd* harmonics (including the fundamental) and the disappearance of all *even* harmonics.

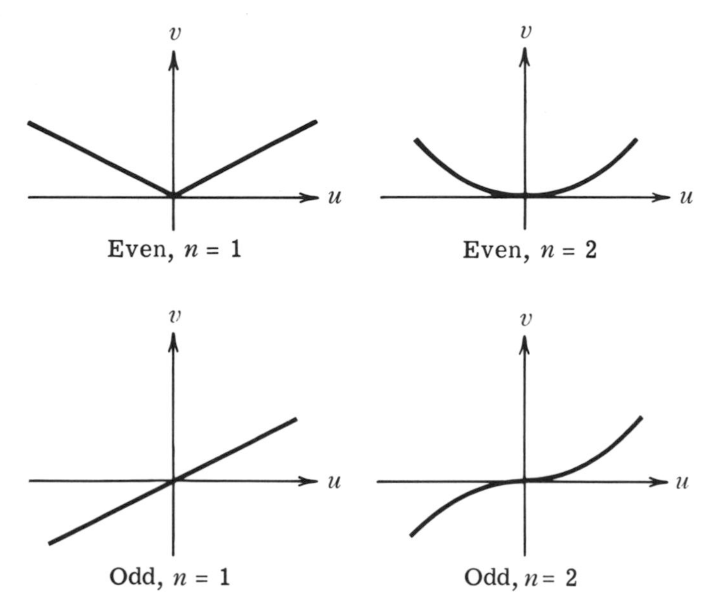


*Figure 5-3.* The even and odd parts of $v(u)$ for the half-wave linear ($n = 1$) and quadratic ($n = 2$) rectifiers. These are also (half) the output of the corresponding full-wave $n$th-power devices of even and of odd symmetry.

When $n$ is an even integer, the full-wave rectifier thus yields only $\frac{1}{2}n + 1$ nonvanishing terms of (5-2), for $\cos^n \theta$ can be expressed in terms of $\cos n\theta$, $\cos (n - 2)\theta$, $\cos (n - 4)\theta$, . . . , 1; for example,

$$\cos^4 \theta = \tfrac{1}{8} \cos 4\theta + \tfrac{1}{2} \cos 2\theta + \tfrac{3}{8}.$$

Similarly, when $n$ is an odd integer, the power-law device of odd symmetry gives $v = u^n$ for all $u$ and yields only $\frac{1}{2}(n + 1)$ nonvanishing terms of (5-2)—the odd harmonics up to the $n$th.

Since (5-7) is half the sum of the outputs of the two associated full-wave devices, it produces all of the harmonics generated by either of them and is thus the most useful power-law device to study. This resolution of (5-7) into its even and odd parts shows us why the output of a half-wave rectifier contains only a finite number of odd or even harmonics when $n$ is, respectively, an odd or an even integer.

**The $m$th-Harmonic Output Signal and Noise.** We now suppose that the input $u(t)$ of the half-wave power-law rectifier is the sum of a signal

$A(t) \cos [2\pi Ft + \alpha(t)]$ plus narrow-band gaussian noise of rms value $\sigma$. The $m$th-harmonic output, by (5-2) and (5-8), is

$$\begin{aligned} v_m(a) \cos [2m\pi Ft + m\alpha + m(\phi - \alpha)] \\ = b_m(n)a^n \cos m(\phi - \alpha) \cos (2m\pi Ft + m\alpha) \\ - b_m(n)a^n \sin m(\phi - \alpha) \sin (2m\pi Ft + m\alpha). \qquad (5\text{-}10) \end{aligned}$$

Using (4-30) and (4-34), modified slightly to take account of the phase $\alpha$ of the input signal, we find that the $m$th-harmonic output signal, the conditional expectation of (5-10) for given values of the input-signal amplitude and phase, is

$$b_m(n)(\tfrac{1}{2}m + \tfrac{1}{2}n)!(A^m/m!)(\sqrt{2}\,\sigma)^{n-m}{}_1F_1(\tfrac{1}{2}m - \tfrac{1}{2}n; m + 1; -A^2/2\sigma^2) \cos (2m\pi Ft + m\alpha). \qquad (5\text{-}11)$$

In the low-frequency case ($m = 0$), an additional factor $\frac{1}{2}$ must be included in both (5-10) and (5-11), and from the latter we must subtract the d-c component

$$\tfrac{1}{2}b_0(n)(\tfrac{1}{2}n)!(\sqrt{2}\,\sigma)^n \operatorname{E}\{{}_1F_1(-\tfrac{1}{2}n;1;-A^2/2\sigma^2)\}, \qquad (5\text{-}12)$$

which is the average value of half of (5-11) with $m = 0$. The expectation here is over the distribution of the signal amplitude $A$.

The $m$th-harmonic output-signal power, the mean-squared value of (5-11), is

$$2^{n-m-1}b_m^2(n)[(\tfrac{1}{2}m + \tfrac{1}{2}n)!/m!]^2\sigma^{2n-2m} \operatorname{E}\{A^{2m}{}_1F_1^2(\tfrac{1}{2}m - \tfrac{1}{2}n;m + 1;-A^2/2\sigma^2)\} \qquad (5\text{-}13)$$

for $m \geq 1$, and for $m = 0$ it is half of (5-13) minus the square of (5-12).

The *total* $m$th-harmonic output power is the mean-squared value of $b_m(n)a^n \cos (2m\pi Ft + m\phi)$. Because of the ergodicity of the signal, its phase has a uniform distribution, and the square of the cosine simply contributes a factor $\frac{1}{2}$. Thus, using (4-35) to evaluate $\operatorname{E}\{a^{2n}\}$, we find that the total $m$th-harmonic output power is

$$2^{n-1}b_m^2(n)n!\sigma^{2n} \operatorname{E}\{{}_1F_1(-n;1;-A^2/2\sigma^2)\} \qquad (5\text{-}14)$$

for $m \geq 1$ and is half this much for $m = 0$.

Since the $m$th-harmonic output signal (including distortion) is orthogonal to the $m$th-harmonic output noise, the average noise power is simply the difference between the total output power (5-14) and the output-signal power (including direct current for $m = 0$) (5-13). [The low-frequency output-noise power is *half* the difference between (5-14) and (5-13).] Thus, setting

$$r = \frac{A^2}{2\sigma^2},$$

the input signal-to-noise ratio when the input-signal amplitude is $A$, we can express the $m$th-harmonic output signal-to-noise ratio (5-3) as

$$R = \frac{[(\frac{1}{2}m + \frac{1}{2}n)!]^2 \operatorname{E}\{r^m {}_1F_1{}^2(\frac{1}{2}m - \frac{1}{2}n;m+1;-r)\}}{(m!)^2 n! \operatorname{E}\{{}_1F_1(-n;1;-r)\} - [(\frac{1}{2}m + \frac{1}{2}n)!]^2 \operatorname{E}\{r^m {}_1F_1{}^2(\frac{1}{2}m - \frac{1}{2}n;m+1;-r)\}}. \tag{5-15}$$

These expectations are to be taken over the distribution of the input-signal amplitude $A$, upon which $r$ depends. As we have noted, when $n$ is an integer and $m = n + 2, n + 4, \ldots$, we have $b_m(n) = 0$ and there is no output in these harmonic zones; (5-15) is then merely the limit of the output signal-to-noise ratio as $n$ approaches the integer. The low-frequency ($m = 0$) output signal-to-noise ratio (5-5) is similarly

$$R = \frac{\operatorname{E}\{{}_1F_1{}^2(-\frac{1}{2}n;1;-r)\} - [\operatorname{E}\{{}_1F_1(-\frac{1}{2}n;1;-r)\}]^2}{n![(\frac{1}{2}n)!]^{-2} \operatorname{E}\{{}_1F_1(-n;1;-r)\} - \operatorname{E}\{{}_1F_1{}^2(-\frac{1}{2}n;1;-r)\}}. \tag{5-16}$$

If the signal amplitude $A$ is constant, (5-16) is zero and no averaging is needed in (5-15). Using (4-33) and the result of Prob. 4-5, we find that in this case (5-15) becomes

$$R = \frac{2r}{m^2 + n^2} \tag{5-17}$$

for large $r$. Using (4-31), we find that for small $r$ (5-15) becomes

$$R = \frac{[(\frac{1}{2}m + \frac{1}{2}n)!]^2}{(m!)^2 n!} r^m, \tag{5-18}$$

again with $m \geq 1$.

As $m$ increases, (5-18) decreases. However, for a given value of $m$, (5-18) is largest when $n = m$. In this sense, an $m$th-power device is optimum for generating the $m$th harmonic.

If $A$ is not constant, (5-15) becomes the expectation of (5-18) for small $r$, and

$$R = \frac{\operatorname{E}\{A^{2n}\}}{(m^2 + n^2)\sigma^2 \operatorname{E}\{A^{2n-2}\}} \tag{5-19}$$

for large $r$, while (5-16) becomes

$$R = \frac{\operatorname{E}\{A^{2n}\} - [\operatorname{E}\{A^n\}]^2}{n^2\sigma^2 \operatorname{E}\{A^{2n-2}\}} \tag{5-20}$$

for large input signal-to-noise ratios, and

$$R = \frac{n^2}{\binom{n}{\frac{1}{2}n} - 1} \, \frac{\operatorname{E}\{A^4\} - [\operatorname{E}\{A^2\}]^2}{16\sigma^4} \tag{5-21}$$

when the signal is weak. Here we see that the low-frequency output signal-to-noise ratio is proportional to the square of the input-signal power when the latter is small, regardless of the value of $n$. (An ideal limiter, with $n = 0$, yields no low-frequency output signal or noise.) This effect is known as suppression of the signal by the noise and results in a great decrease in signal-to-noise ratio from input to output. The first factor in (5-21) is greatest for $n = 2$; that is, among all power-law detectors, the square-law detector yields the greatest output signal-to-noise ratio at low signal-to-noise ratios.

**Problem 5-8.** For the case of a strong signal of fixed amplitude $A$ and phase $\alpha$, using the approximations $a = A + x$ and $\phi = \alpha + y/A$ (see Fig. 4-1), resolve the $m$th-harmonic output (5-10) into its signal and noise components. Show that the output-noise spectrum has the same shape as the input-noise spectrum if the latter is symmetric about the signal frequency, and obtain (5-17).

**Problem 5-9.** Show that (5-15) becomes (5-18) or (5-19) and that (5-16) becomes (5-20) or (5-21) when the input signal-to-noise ratio is very large or very small.

**Problem 5-10.** By considering a nonlinear device with $v(u)$ a polynomial in $u$ whose input $u$ is the sum of narrow-band noise plus a narrow-band signal, show that the low-frequency output signal-to-noise ratio is quite generally proportional to the square of the input signal-to-noise ratio when the latter is small, the low-frequency output signal then being proportional to the square of the input-signal amplitude and the output noise independent of it. How will the $m$th-harmonic output signal-to-noise ratio behave when the input signal is weak?

**The Ideal Limiter** ($n = 0$). For the case of a constant-amplitude phase- or frequency-modulated signal passed through an ideal limiter along with narrow-band gaussian noise to remove the amplitude fluctuations of their sum, we set $m = 1$ and $n = 0$ in (5-15) in order to find the first-harmonic output signal-to-noise ratio. Since ${}_1F_1(0;1;-r) = 1$ by (4-31) and[1] ${}_1F_1(\frac{1}{2};2;-r) = [I_0(\frac{1}{2}r) + I_1(\frac{1}{2}r)] \exp(-\frac{1}{2}r)$, and since $r$ is constant in this case, (5-15) becomes

$$R = \frac{\pi r[I_0(\frac{1}{2}r) + I_1(\frac{1}{2}r)]^2}{4e^r - \pi r[I_0(\frac{1}{2}r) + I_1(\frac{1}{2}r)]^2}. \tag{5-22}$$

[1] D. Middleton, "An Introduction to Statistical Communication Theory," appendix 1.2, pp. 1073–1076, McGraw-Hill Book Company, New York, 1960; and S. O. Rice in Nelson Wax (ed.), "Noise and Stochastic Processes," appendix 4B, Dover Publications, Inc., New York, 1954.

For small $r$, this is evidently $R = \frac{1}{4}\pi r$, in agreement with (5-18), and for large $r$ it becomes $R = 2r$ by (5-17). Thus the band-pass limiter alters its input signal-to-noise ratio by something between $10 \log_{10} \frac{1}{4}\pi = -1.05$ db and $10 \log_{10} 2 = +3$ db, the latter increase being due to the limiter's eliminating the $x$ component of the input noise, which is in phase with the input signal and causes only amplitude fluctuations of the input when the signal is strong.

**The Linear Rectifier** ($n = 1$). Setting $m = n = 1$ in (5-15), we find that the output signal-to-noise ratio of a band-pass linear rectifier is just its average input signal-to-noise ratio, for the even part of $v(u)$ here produces no odd harmonics and its odd part is simply $\frac{1}{2}u$, that is, the first-harmonic output is exactly half the input.

To find the low-frequency output signal-to-noise ratio of a linear detector, we set $n = 1$ in (5-16). Since ${}_1F_1(-1;1;-r) = 1 + r$ and ${}_1F_1(-\frac{1}{2};1;-r) = [(1 + r)I_0(\frac{1}{2}r) + rI_1(\frac{1}{2}r)] \exp(-\frac{1}{2}r)$, we get

$$R = \frac{\mathrm{E}\,\{e^{-r}[(1 + r)I_0(\frac{1}{2}r) + rI_1(\frac{1}{2}r)]^2\} - \mathrm{E}^2\,\{e^{-\frac{1}{2}r}[(1 + r)I_0(\frac{1}{2}r) + rI_1(\frac{1}{2}r)]\}}{\mathrm{E}\,\{(4/\pi)(1 + r) - e^{-r}[(1 + r)I_0(\frac{1}{2}r) + rI_1(\frac{1}{2}r)]^2\}}, \tag{5-23}$$

which generally requires numerical integration to average over the distribution of the input-signal amplitude $A$ with $r = A^2/2\sigma^2$. (If $A$ is constant, there is no low-frequency output signal, and $R = 0$.)

When the signal is very strong or very weak, however, (5-23) simplifies considerably, and analytic methods can be used. In the former case, it reduces to (5-20), which for $n = 1$ is

$$R = \mathrm{E}\,\{A^2/\sigma^2\} - [\mathrm{E}\,\{A/\sigma\}]^2. \tag{5-24}$$

When the signal is weak, (5-23) becomes (5-21), whose first factor has the value $\pi/(4 - \pi)$ for $n = 1$.

**Problem 5-11.** Show that the low-frequency output signal-to-noise ratio of a linear detector whose input is narrow-band gaussian noise plus a strong or a weak signal amplitude modulated by a sinusoid as in Prob. 5-7 is $R = 2m^2r/(1 + m^2)$ or $R = (4 - \pi)^{-1}\pi m^2r^2(1 + \frac{1}{8}m^2)/(1 + m^2)^2$, respectively, where $r$ is the average input signal-to-noise ratio and $m$ is the *rms modulation index*, i.e., the ratio of the standard deviation of the input-signal amplitude to its mean. Notice that, when the input signal is strong, the output signal reproduces the variations of its amplitude without distortion, but when it is weak, as we saw in Prob. 5-10, the output signal varies as the square of the input amplitude. To the latter case, therefore, apply the result of Prob. 5-7 to obtain the ratio of the undistorted output-signal power to the output-noise power. Also treat the case of gaussian modulation.

**The Square-Law Detector** ($n = 2$). Since

$${}_1F_1(-2;1;-r) = 1 + 2r + \tfrac{1}{2}r^2,$$

the low-frequency output signal-to-noise ratio of a square-law detector is (5-16) with $n = 2$,

$$\begin{aligned} R &= \frac{\mathrm{E}\,\{r^2\} - [\mathrm{E}\,\{r\}]^2}{1 + 2\,\mathrm{E}\,\{r\}} \\ &= \frac{1}{4\sigma^2}\,\frac{\mathrm{E}\,\{A^4\} - [\mathrm{E}\,\{A^2\}]^2}{\sigma^2 + \mathrm{E}\,\{A^2\}}, \end{aligned} \tag{5-25}$$

which is nearly as simple as the two approximations (5-20) and (5-21) with $n = 2$, and is likewise easily evaluated. (See Prob. 5-7.)

The case of the square-law detector is the most tractable because we may suppose it to be a full-wave rectifier with $v = u^2$ for all $u$ insofar as low-frequency and even-harmonic outputs are concerned, and hence (5-2) becomes simply

$$\begin{aligned} v &= a^2(t) \cos^2 [2\pi Ft + \phi(t)] \\ &= \tfrac{1}{2}a^2 + \tfrac{1}{2}a^2 \cos (4\pi Ft + 2\phi). \end{aligned}$$

Thus, the low-frequency output is $\frac{1}{2}a^2$, as in Prob. 5-7, and the remainder of the output, $\frac{1}{2}a^2 \cos (4\pi Ft + 2\phi)$, falls in the second-harmonic zone.

With the sum of a signal of amplitude $A$ plus narrow-band gaussian noise of power $\sigma^2$ as input, the low-frequency output $\frac{1}{2}a^2$ has the conditional expectation (conditioned on $A$) $\frac{1}{2}A^2 + \sigma^2$ by Prob. 4-3. Hence the d-c output component is $\frac{1}{2}\,\mathrm{E}\,\{A^2\} + \sigma^2$, the output signal is

$$\tfrac{1}{2}A^2 - \tfrac{1}{2}\,\mathrm{E}\,\{A^2\},$$

the output-signal power is

$$\mathrm{E}\,\{[\tfrac{1}{2}A^2 - \tfrac{1}{2}\,\mathrm{E}\,\{A^2\}]^2\} = \tfrac{1}{4}\,\mathrm{E}\,\{A^4\} - \tfrac{1}{4}[\mathrm{E}\,\{A^2\}]^2,$$

the output noise is $\frac{1}{2}a^2 - \frac{1}{2}A^2 - \sigma^2$, and the output-noise power is the difference between the total output power

$$\tfrac{1}{4}\,\mathrm{E}\,\{a^4\} = \tfrac{1}{4}\,\mathrm{E}\,\{A^4\} + 2\sigma^2\,\mathrm{E}\,\{A^2\} + 2\sigma^4$$

and the sum of the output-signal and d-c powers, $\mathrm{E}\,\{(\frac{1}{2}A^2 + \sigma^2)^2\}$, namely, $\sigma^4 + \sigma^2\,\mathrm{E}\,\{A^2\}$, which again gives us (5-25). For the second-harmonic zone, (5-15) with $m = n = 2$ gives

$$\begin{aligned} R &= \frac{\frac{1}{2}\,\mathrm{E}\,\{r^2\}}{1 + 2\,\mathrm{E}\,\{r\}} \\ &= \frac{1}{8\sigma^2}\,\frac{\mathrm{E}\,\{A^4\}}{\sigma^2 + \mathrm{E}\,\{A^2\}}. \end{aligned} \tag{5-26}$$

**Problem 5-12.** Show that, if a full-wave quadratic rectifier with $v = u^2$ is fed a signal $A(t) \cos [2\pi Ft + \alpha(t)]$ plus independent, zero-mean noise of any statistics whatever, the output signal is $\frac{1}{2}A^2(t) - \frac{1}{2} \mathrm{E}\{A^2\} + \frac{1}{2}A^2(t) \cos [4\pi Ft + 2\alpha(t)]$, the sum of the first two terms being the low-frequency output signal and the third term being the second-harmonic.

**Problem 5-13.** Using the result of Prob. 4-1, show that the $m$th-harmonic output of an $n$th-power nonlinearity with narrow-band gaussian input of rms bandwidth $\rho$ has rms bandwidth[1] $\rho \sqrt{(m^2 + n^2)/2n}$. Note that this reduces to the result of Prob. 4-9 when $m = 0$. [HINT: Notice that (Prob. 4-10) $p(\phi'|a)$ is a zero-mean, normal distribution with variance $4\pi^2\rho^2\sigma^2/a^2$; recall that, by (4-37), $a'$ is independent of $a$ and $\phi'$ and has mean-squared value $4\pi^2\rho^2\sigma^2$; and obtain the required moments of the distribution of $a$ from (4-35) with $A = 0$.]

**Problem 5-14.** Notice that, for any given $m$, the bandwidth obtained in the preceding problem is least when $n = m$. That is, among all power-law nonlinearities, those with exponent $m$ give the smallest $m$th-harmonic rms output bandwidth $\sqrt{m}\,\rho$ for a given narrow-band gaussian input of rms bandwidth $\rho$. Show that such nonlinearities still uniquely yield the minimum bandwidth when we allow *any* nonlinearity or, equivalently, any $v_m(a)$ at all. [HINT: To show that $\rho\sigma \sqrt{\mathrm{E}\{v'^2 + m^2v^2/a^2\}/\mathrm{E}\{v^2\}} \geq \sqrt{m}\,\rho$, where E denotes the average over the Rayleigh distribution of $a$, the subscript on $v_m$ has been suppressed for convenience, and $v'(a) = dv/da$, replace $m$ by the value $m = \mathrm{E}\{v^2\}/2\sigma^2\,\mathrm{E}\{v^2/a^2\}$ (not necessarily an integer) that minimizes $\mathrm{E}\{v'^2 + m^2v^2/a^2 - mv^2/\sigma^2\}$, and make use of Schwarz's inequality along with integration by parts.]

**Problem 5-15.** Generalize the result of Prob. 5-13 to include with the input noise a signal of constant amplitude $A$ whose frequency coincides with the centroid of the input-noise spectrum, obtaining

$$\rho \sqrt{\frac{m^2 + n^2}{2n}} \sqrt{\frac{{}_1F_1(-n + 1;1;-A^2/2\sigma^2)}{{}_1F_1(-n;1;-A^2/2\sigma^2)}}$$

for the rms bandwidth of the $m$th-harmonic output signal and noise with $\sigma$ the rms input noise. Show that, to obtain the rms bandwidth of the $m$th-harmonic output noise alone, the second denominator must be reduced by $[(\frac{1}{2}m + \frac{1}{2}n)!/m!]^2(A^2/2\sigma^2)^m {}_1F_1{}^2(\frac{1}{2}m - \frac{1}{2}n;m + 1;-A^2/2\sigma^2)/n!$.

## ☐ OUTPUT SPECTRUM

**Full-Wave Square-Law Rectifier.** It is often important to know not only the output signal and noise power within a given harmonic zone

[1] Norman M. Abramson, Nonlinear Transformations of Random Processes, to be published in *IEEE Trans. Inform. Theory.*

but also the spectral distribution of this power, for further filtering can frequently be used to remove much of the output noise without affecting the desired output signal. In the case of the full-wave quadratic rectifier we can easily determine the output spectrum by means of the Wiener-Khinchin theorem.

Thus we suppose that its input $u(t)$ is the sum of a signal $s(t)$ plus independent, zero-mean noise $n(t)$. The correlation function of its output

$$\begin{aligned} v &= u^2(t) \\ &= [s(t) + n(t)]^2 \end{aligned}$$

is therefore

$$\begin{aligned} \psi_v(\tau) &= \mathrm{E}\,\{[s(t) + n(t)]^2[s(t+\tau) + n(t+\tau)]^2\} \\ &= \mathrm{E}\,\{s^2(t)s^2(t+\tau)\} + \mathrm{E}\,\{n^2(t)n^2(t+\tau)\} + 4\psi_s(\tau)\psi_n(\tau) \\ &\qquad + 2\psi_s(0)\psi_n(0), \end{aligned} \tag{5-27}$$

where $\psi_s(\tau) = \mathrm{E}\,\{s(t)s(t+\tau)\}$ and $\psi_n(\tau) = \mathrm{E}\,\{n(t)n(t+\tau)\}$ are the correlation functions of the input signal and the input noise, respectively, and $\psi_s(0)$ and $\psi_n(0) = \sigma^2$ are their mean-squared values.

If, for example, the signal is a sinusoid $s(t) = A\cos(2\pi Ft + \alpha)$ of fixed amplitude $A$ and constant phase $\alpha$, by (3-23) its correlation function is $\psi_s(\tau) = \frac{1}{2}A^2\cos 2\pi F\tau$ and that of its square is

$$\mathrm{E}\,\{s^2(t)s^2(t+\tau)\} = \tfrac{1}{4}A^4 + \tfrac{1}{8}A^4\cos 4\pi F\tau;$$

making use of the result of Prob. 1-15, we see that, if the noise is gaussian, $\mathrm{E}\,\{n^2(t)n^2(t+\tau)\} = 2\psi_n^2(\tau) + \psi_n^2(0)$. Substituting these values into (5-27), we get

$$\begin{aligned} \psi_v(\tau) &= \tfrac{1}{8}A^4\cos 4\pi F\tau + 2\psi_n^2(\tau) + 2A^2\psi_n(\tau)\cos 2\pi F\tau \\ &\qquad + (\tfrac{1}{2}A^2 + \sigma^2)^2. \end{aligned} \tag{5-28}$$

The output power spectral density is the Fourier transform of (5-28),

$$\begin{aligned} \Psi_v(f) &= \tfrac{1}{16}A^4[\delta(f - 2F) + \delta(f + 2F)] + 2\Psi_n(f) \star \Psi_n(f) \\ &\qquad + A^2[\Psi_n(f - F) + \Psi_n(f + F)] + (\tfrac{1}{2}A^2 + \sigma^2)^2\delta(f). \end{aligned} \tag{5-29}$$

The first term of (5-29) represents the second harmonic of the input signal. The second term, twice the convolution of the input-noise spectrum with itself, represents noise–noise beats, i.e., the sum and difference frequencies that result from the cross products when the sum of the Fourier components of the noise is squared. The third term similarly represents the beats between the input signal and the input noise, which here simply shift the input-noise spectrum by $\pm F$. The last term of (5-29) represents the d-c output component, $\frac{1}{2}A^2 + \sigma^2$, which is the mean-squared input.

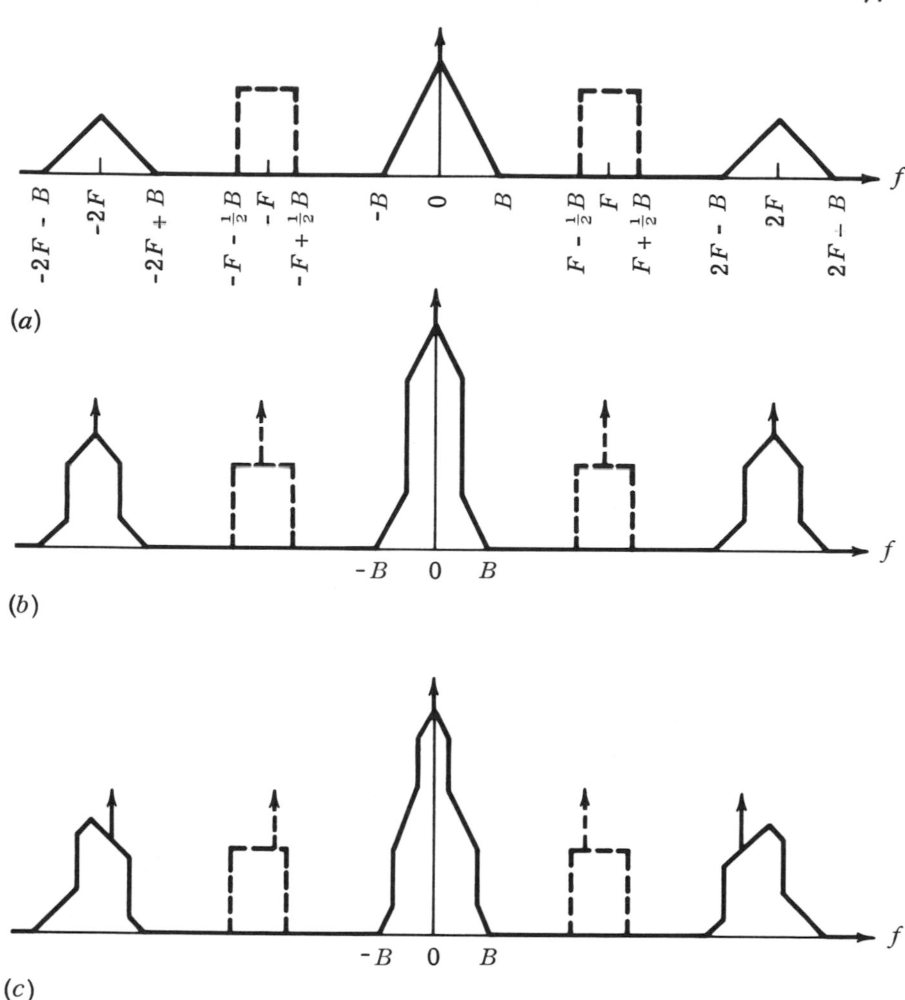


**Figure 5-4.** Output power spectral density, shown solid, of a square-law device whose input is gaussian noise with a rectangular power spectral density, shown dashed, plus a sinusoid, also shown dashed, (*a*) of zero amplitude, (*b*) in the center of the noise spectrum, and (*c*) eccentric. Arrows indicate the spectral lines. Note that the input need not be narrow-band.

Figure 5-4 shows this spectrum for the case of a rectangular input-noise spectrum; here it consists of discrete lines representing the d-c and second-harmonic signal output, triangular parts representing noise–noise beats, and rectangular parts representing signal–noise beats. Notice that in this case an output filter passing only frequencies from $-b$ to $b$, with $b$ very small compared to the bandwidth $B$ of the input noise, will pass a fraction $2b/B$ of the low-frequency output noise regardless of the amplitude of the input signal or its frequency (provided that it lies inside the input-noise band), for near $f = 0$ both the triangular and rectangular parts of $\Psi_v(f)$ equal twice their average value over the range from $-B$ to $B$, and a narrow vertical strip of width $2b$ in the middle of Fig. 5-1 will include $2b/B$ of the central area. Thus, provided that it passes the entire output signal, such a filter will improve the low-frequency output signal-to-noise ratio (5-25) by the factor $B/2b$.

**Problem 5-16.** Determine all of the dimensions of Fig. 5-4c, calculate the areas representing the low-frequency and second-harmonic noise, and verify that they are correct, viz., both $A^2\sigma^2 + \sigma^4$.

**Problem 5-17.** Find the output spectrum of a square-law device having a gaussian input with power spectral density $\Psi_n(f) = \Sigma_{\pm}\pi^{-1}QF\sigma^2/[F^2 + 4Q^2(f \pm F)^2]$, where the summation includes a term with each sign. This input-noise spectrum can result from passing wider-band noise through a high-$Q$ simple resonant circuit.

**Problem 5-18.** Find the output spectrum of a square-law device having a gaussian input with power spectral density $\Psi_n(f) = \Sigma_{\pm}(8\pi)^{-\frac{1}{2}}(\sigma^2/B)\exp[-(f \pm F)^2/2B^2]$. Such a spectrum can result from passing wider-band noise through a large number of filters peaked at the same frequency $F$.

**Problem 5-19.** With only gaussian noise as input, (5-29) becomes $\Psi_v(f) = 2\Psi_n(f) \star \Psi_n(f) + \sigma^4\delta(f)$. Into this formula substitute the input spectrum $\frac{1}{4}A^2[\delta(f - F) + \delta(f + F)] + \Psi_n(f)$, which describes the signal-and-noise input upon which (5-29) is based, and notice that the result agrees with (5-29) except for the doubling of the terms containing $A^4$. This approach makes the sinusoid a part of the gaussian input noise. As a result it becomes a sinusoidal gaussian random process with a constant, Rayleigh-distributed amplitude having mean-squared value $A^2$ instead of the prescribed amplitude $A$. Hence, by Prob. 1-10, the fourth moment of the amplitude is $2A^4$ instead of $A^4$.

**Problem 5-20.** Determine the output spectrum of a square-law device whose input consists of two sinusoids of incommensurable frequencies plus gaussian noise with a rectangular spectrum as in Fig. 5-4. Use the method suggested by Prob. 5-19, that is, include the sinusoids in the noise spectrum and then halve all terms involving the fourth power of the amplitude of either of them.

**Problem 5-21.** As in Prob. 5-7, a high-frequency sinusoidal carrier is amplitude modulated by a gaussian random process, its spectrum being flat from $-\frac{1}{2}B$ to $\frac{1}{2}B$ and zero outside this range. The modulated carrier is fed to a square-law detector along with gaussian noise having a rectangular spectrum that occupies the same band of frequencies, and the detector output is fed to an ideal low-pass filter of cutoff frequency $\frac{1}{2}B$. Regarding all distortion as noise, determine the output signal-to-noise ratio and indicate the shape of the output-noise spectrum.

**Polynomial Nonlinearities.** The foregoing approach is useful not only for the square-law device but also for any nonlinearity whose output $v$ can be expressed or approximated as a polynomial in its input $u$,

$$v = c_0 + c_1u + c_2u^2 + \cdots + c_ku^k. \tag{5-30}$$

If $u$ is regarded as the sum of its Fourier components, the term $c_mu^m$ involves their $m$-fold products, which, by the trigonometric identity for the product of two cosines, yield all sums and differences among $m$ input frequencies, i.e., $m$th-order beats.

When the nonlinearity is used to demodulate a narrow-band input, only the low-frequency portion of the output is of interest, and it results entirely from the even-power terms of (5-30), since only beats of even order can produce low frequencies. The constant term $c_0$ represents only direct current and is of no interest. The output signal resulting from the quadratic term $c_2u^2$, as we have seen, is proportional to the square of the input signal, since it is due to signal–signal beats. Likewise, the output signal resulting from higher-power terms is proportional to the square of the input signal when the latter is weak, since the dominant output-signal component is then due to beats involving the signal just twice. When the input signal is weak, the output-noise power is nearly independent of it, and hence, when the input signal-to-noise ratio $r$ is small, the output signal-to-noise ratio is proportional to its square, regardless of the form (5-30) of the detection characteristic. (See Prob. 5-10.) Similarly, when $r$ is large, by (5-20) each even-power term of (5-30) yields an output noise-to-signal ratio proportional to $1/r = 2\sigma^2/A^2$, and so the output-noise power varies as $\sigma^2$ for $\sigma \ll A$. Hence, the output signal-to-noise ratio is proportional to $r$ for large $r$—again regardless of the detection characteristic.

**Problem 5-22.** Suppose that (5-30), with $k = 4$, is the unintentionally nonlinear characteristic of a device intended to amplify a narrow-band gaussian random process of rms value $\sigma$ having a rectangular spectrum. Show that at the center of this spectrum the ratio of the undistorted output-signal power spectral density to the output-distortion power spectral density has its minimum value, $\frac{8}{3}(1 + c_1/3c_3\sigma^2)^2$.

**The Linear Rectifier with Gaussian Input.**[1] The direct method which we used to find the output correlation function, and thus the output

[1] Rice, *op. cit.*, sec. 4.7.

power spectrum, in the case of the square-law device can be used not only for polynomial nonlinearities but also for certain other nonlinearities, provided that the bivariate probability density function $p(u_1, u_2)$ of the inputs $u_1 = u(t)$ and $u_2 = u(t + \tau)$ is simple enough to permit evaluation of the resulting double integral,

$$\begin{aligned}\psi_v(\tau) &= \mathrm{E}\,\{v(u_1)v(u_2)\} \\ &= \iint_{-\infty}^{\infty} v(u_1)v(u_2)p(u_1, u_2)\,du_1\,du_2. \end{aligned} \tag{5-31}$$

This is the case, for example, with a half-wave linear rectifier whose input is just zero-mean gaussian noise. Here $v(u) = u$ for $u \geq 0$ and $v(u) = 0$ for $u \leq 0$, and $p(u_1,u_2)$ is given by (1-21) with

$$\mu_1 = \mu_2 = 0,\ M_{11} = M_{22} = \sigma^2 = \psi(0),$$

and $M_{12} = M_{21} = \psi(\tau)$, the correlation function of the input noise. Substituting these values into (5-31), which permit changing the lower limits to zero, making the change of variables suggested in Prob. 1-7, and integrating first with respect to $r$ and then with respect to $\theta$, we get

$$\begin{aligned}\psi_v(\tau) &= \frac{1}{2\pi}\sqrt{\sigma^4 - \psi^2} + \frac{1}{2\pi}\psi \arccos - \frac{\psi}{\sigma^2} \\ &= \tfrac{1}{4}\psi + \frac{\sigma^2}{2\pi}\left(1 + \frac{1}{2}\frac{\psi^2}{\sigma^4} + \frac{1}{2\cdot 4}\frac{\psi^4}{3\sigma^8} + \frac{1\cdot 3}{2\cdot 4\cdot 6}\frac{\psi^6}{5\sigma^{12}}\right. \\ &\qquad \left. + \frac{1\cdot 3\cdot 5}{2\cdot 4\cdot 6\cdot 8}\frac{\psi^8}{7\sigma^{16}} + \cdots\right) \end{aligned} \tag{5-32}$$

after expansion as a power series in $\psi = \psi(\tau)$, the arccosine being between 0 and $\pi$.

The term $\frac{1}{4}\psi$ results from the odd part of the rectifier characteristic, $\frac{1}{2}u$, which simply yields an output component equal to half the input noise. The term $\sigma^2/2\pi$ is the only one that does not vanish as $\tau$ becomes infinite and $\psi(\tau)$ goes to zero; by comparison with (3-23) we see that it represents the d-c component of the output, which has the value $\sigma/\sqrt{2\pi}$. The term $\psi^2/4\pi\sigma^2$ represents the same sort of beats between input components that arise in the case of a square-law device, and it similarly contributes to the output spectrum $1/4\pi\sigma^2$ times the convolution of the input-noise spectrum with itself. The terms involving higher powers of $\psi$ represent higher-order beats and, on Fourier transformation, contribute to the output spectrum higher-order convolutions of the input spectrum with itself.

Setting $\tau = 0$ in (5-32), we find the total output power $\frac{1}{2}\sigma^2$ expressed as the sum of the contributions of these separate terms, namely, $\sigma^2/4 + \sigma^2/2\pi + \sigma^2/4\pi + \sigma^2/48\pi + \sigma^2/160\pi + 5\sigma^2/1792\pi + \cdots$. Hence the first three terms of (5-32), which give the first three terms of

$$\Psi_v(f) = \tfrac{1}{4}\Psi(f) + \frac{\sigma^2}{2\pi}\,\delta(f) + \frac{1}{4\pi\sigma^2}\,\Psi(f) \star \Psi(f) + \cdots, \qquad (5\text{-}33)$$

account for $\frac{1}{2} + 3/2\pi = 97.7$ percent of the total output power. However, the accurate evaluation of $\Psi_v(f)$ for any particular $f$ may require many more terms of the series. With a rectangular input spectrum, these three terms of $\Psi_v(f)$ give an output spectrum like that shown in Fig. 5-4*a* but with the dashed portions made solid so as to include the first term of (5-33).

**Problem 5-23.** Use (5-31) to find the correlation function of the output of a half-wave square-law rectifier with zero-mean gaussian input. Why does it differ from (5-28) with $A = 0$?

**Characteristic-Function Method.**† In order to be able to include a signal

$$s(t) = A(t) \cos [2\pi F t + \alpha(t)],$$

along with the zero-mean gaussian noise, we use Rice's characteristic-function method[1] to find the correlation function of the output and, from it, the output spectrum. Here the nonlinearity $v(u)$ is expressed as a Fourier transform or, more generally, as

$$v(u) = \int_C V(z) e^{iuz}\, dz, \qquad (5\text{-}34)$$

where $C$ is an appropriate path of integration. For example, the half-wave $n$th-power rectifier (5-7) is given by

$$V(z) = \frac{n!}{2\pi(iz)^{n+1}}, \qquad (5\text{-}35)$$

with $C$ the real axis from $-\infty$ to $\infty$ with a downward indentation at the origin; again $n$ need not be an integer.

Substituting (5-34) into (5-31) and interchanging the order of integration and averaging, we get

$$\begin{aligned}\psi_v(\tau) &= \int_C \int_C V(z_1)V(z_2)\, \mathrm{E}\,\{e^{iu_1z_1 + iu_2z_2}\}\, dz_1\, dz_2 \\ &= \int_C \int_C V(z_1)V(z_2)X(z_1,z_2)\, dz_1\, dz_2, \qquad (5\text{-}36)\end{aligned}$$

† May be omitted on the first reading.

[1] Rice, *op. cit.*, sec. 4.8ff. See also Middleton, *op. cit.*, chaps. 5 and 13; Ralph Deutsch, "Nonlinear Transformations of Random Processes," chap. 2, Prentice-Hall, Inc., Englewood Cliffs, N.J., 1962.

in which $X(z_1,z_2)$ denotes the characteristic function of the bivariate distribution of $u_1 = u(t)$ and $u_2 = u(t + \tau)$. With the input noise statistically independent of the input signal, the characteristic function of their sum $u(t)$ is the product of their separate characteristic functions. That of the noise is given by (1-26), namely,

$$\exp\left(-\tfrac{1}{2}\sigma^2 z_1^2 - \psi z_1 z_2 - \tfrac{1}{2}\sigma^2 z_2^2\right). \tag{5-37}$$

To find the characteristic function of the signal, we use (4-27) to obtain the Fourier-series expansion

$$e^{izA\cos(2\pi Ft+\alpha)} = \sum_{m=0}^{\infty} \epsilon_m I_m(iAz) \cos(2m\pi Ft + m\alpha),$$

where $\epsilon_m$, the Neumann factor, is 1 for $m = 0$ and is 2 for all other $m$. Thus the signal's characteristic function

$$X_s(z_1,z_2) = \mathrm{E}\,\{\exp[iz_1A_1 \cos(2\pi Ft + \alpha_1) + iz_2A_2 \cos(2\pi F[t + \tau] + \alpha_2)]\},$$

where $A_1 = A(t)$, $\alpha_1 = \alpha(t)$, $A_2 = A(t + \tau)$, and $\alpha_2 = \alpha(t + \tau)$, can be expressed as the expectation of the product of two such Fourier series. Because of the ergodicity of the signal, $\alpha(t)$ contains a uniformly distributed constant, which results in the vanishing of the expectations of the cross products involving different values of $m$ and in

$$\mathrm{E}\,\{\cos m(2\pi Ft + \alpha_1) \cos m[2\pi F(t + \tau) + \alpha_2]\} = \epsilon_m^{-1}\,\mathrm{E}\,\{\cos m(2\pi F\tau + \alpha_2 - \alpha_1)\}.$$

Thus, since $I_m(iz) = i^m J_m(z)$,

$$X_s(z_1,z_2) = \sum_{m=0}^{\infty} (-1)^m \epsilon_m\,\mathrm{E}\,\{J_m(A_1z_1)J_m(A_2z_2) \cos m(2\pi F\tau + \alpha_2 - \alpha_1)\}. \tag{5-38}$$

When the signal is an unmodulated sinusoid of frequency $F$, we have $\alpha_1 = \alpha_2$, and no averaging is needed in (5-38), $A_1$ and $A_2$ both being its amplitude $A$. Substituting the product of the resulting $X_s(z_1,z_2)$ times (5-37) for $X(z_1,z_2)$ in (5-36) and expanding

$$e^{-\psi z_1 z_2} = \sum_{k=0}^{\infty} (-1)^k \frac{\psi^k z_1{}^k z_2{}^k}{k!}$$

as a power series, we find that the double integral becomes the product of two identical single integrals and

$$\psi_v(\tau) = \sum_{m=0}^{\infty} \sum_{k=0}^{\infty} \frac{\epsilon_m h_{mk}^2}{k!} \psi^k(\tau) \cos 2m\pi F\tau, \tag{5-39}$$

where

$$h_{mk} = i^{m+k} \int_C V(z) z^k J_m(Az) e^{-\sigma^2 z^2/2}\, dz. \tag{5-40}$$

For example, in the case of the half-wave $n$th-power rectifier (5-7), for which $V(z)$ is given by (5-35),[1]

$$h_{mk} = \frac{n!\sigma^{n-m-k} A^m}{2^{(2+m+n-k)/2} m! \left(\dfrac{n-m-k}{2}\right)!} {}_1F_1\left(\frac{m-n+k}{2}; m+1; -\frac{A^2}{2\sigma^2}\right). \tag{5-41}$$

The only terms in (5-39) that do not vanish as $\tau$ grows infinite and $\psi(\tau)$ becomes zero are those with $k = 0$. By comparison with (3-23) we see that the term $h_{00}^2$ represents the power of the d-c output component $h_{00}$, in agreement with (5-12), and the term $2h_{m0}^2 \cos 2m\pi F\tau$ represents the discrete output component which is the $m$th harmonic of the input signal and has amplitude $2h_{m0}$, in agreement with (5-11).

The terms of (5-39) with $k \neq 0$ yield the continuous part of the output spectrum. The general term, after Fourier transformation, contributes the $k$-fold convolution of the input-noise spectrum with itself, convolved with the spectrum of the $m$th harmonic of the input signal. In this sense, it may be described as resulting from beats of order $m + k$ between the noise, included $k$ times, and the signal, included $m$ times, although such signal-signal beats would produce harmonics of order $m - 2$, $m - 4$, . . . , as well as the $m$th. The total output power contributed by the term is found by setting $\tau = 0$ in (5-39), namely, $\epsilon_m h_{mk}^2 \sigma^{2k}/k!$.

**Problem 5-24.** Verify that (5-39) with $A = 0$ becomes (5-32) in the case of the linear rectifier.

**Problem 5-25.** How should (5-39) be modified to include the case of a signal that is amplitude modulated, phase modulated, or both?

**Problem 5-26.** Notice that, with a rectangular input-noise spectrum, the terms of (5-39) with $m + k \leq 2$ yield an output spectrum resembling that shown in Fig. 5-4*a*, *b*, or *c* with the dashed lines made full. Calculate this spectrum for the case of the linear rectifier.

[1] Rice, *op. cit.*, eq. (4.10-5).

**Problem 5-27.** Notice that, when the input to the nonlinear device is zero-mean, gaussian noise alone and $A = 0$, (5-39) becomes $\psi_v(\tau) = \Sigma_0^\infty h_{0k}^2 \psi^k(\tau)/k!$ with $h_{0k}$ equal to the mean value $\mathrm{E}\{v^{(k)}(u)\}$ of the $k$th derivative of $v(u)$, averaged over the normal distribution of the input $u$. Show that, if this input is regarded as a signal, the term $k = 1$ corresponds to the undistorted output signal, which emerges with an effective voltage gain equal to the average slope $\mathrm{E}\{v'(u)\}$ of the nonlinearity. What if the input is the sum of a zero-mean gaussian signal plus independent gaussian noise?

**Problem 5-28.** For Hermite polynomials $\mathrm{He}_k(x) = e^{\frac{1}{2}x^2}(-d/dx)^k e^{-\frac{1}{2}x^2}$ we have $\int_{-\infty}^{\infty} \mathrm{He}_j(x)\mathrm{He}_k(x)e^{-\frac{1}{2}x^2}\,dx = 0$ for $j \neq k$ and $= \sqrt{2\pi}\,k!$ for $j = k$. Using this orthogonality relation, express the nonlinearity in Prob. 5-27 as[1] $v(u) = \Sigma_0^\infty(\sigma^k h_{0k}/k!)\mathrm{He}_k(u/\sigma)$. Show that, when the input $u(t)$ is a zero-mean, gaussian random process with rms value $\sigma$, the crosscorrelation function of any two terms of this series vanishes for all $\tau$ and that each term of this series therefore gives rise to the corresponding term in the series of Prob. 5-27. Thus, the $\mathrm{He}_k(u/\sigma)$ component of $v(u)$ accounts entirely for the $k$th-order beats appearing in the output spectrum, and the $\mathrm{He}_1(u/\sigma) = u/\sigma$ component alone yields undistorted output when the input is gaussian.

**Problem 5-29.** (*a*) Show that the cross-correlation function of the gross output signal and the output noise of any nonlinearity always vanishes identically and that the output-noise spectrum is therefore the difference between the total output spectrum and that of the gross output signal. (*b*) Similarly show that the spectrum of the distortion is the difference between that of the gross output signal and that of the undistorted output signal (5-6), provided only that the input signal is zero-mean and gaussian. [HINT: See Prob. 5-28 or express the bivariate normal distribution of the input signal in the form of (1-13) times (1-3) and make use of (1-14).]

**Problem 5-30.** Show that the correlation functions of the output noise and of the distortion of the error-function limiter of Prob. 5-4 are

$$\frac{2}{\pi}\arcsin\frac{\psi_s(\tau)+\psi_n(\tau)}{\psi_s(0)+\psi_n(0)+\Sigma^2} - \frac{2}{\pi}\arcsin\frac{\psi_s(\tau)}{\psi_s(0)+\psi_n(0)+\Sigma^2}$$

and

$$\frac{2}{\pi}\arcsin\frac{\psi_s(\tau)}{\psi_s(0)+\psi_n(0)+\Sigma^2} - \frac{2}{\pi}\frac{\psi_s(\tau)}{\psi_s(0)+\psi_n(0)+\Sigma^2},$$

respectively, when its input is zero-mean gaussian noise with correlation function $\psi_n(\tau)$ plus a zero-mean gaussian signal having correlation function $\psi_s(\tau)$.

[1] L. L. Campbell, A General Analysis of Post-detection Correlation, *IEEE Trans. Inform. Theory*, **IT-11**(3):409–417 (1965).

# chapter 6 FM DEMODULATION†

Just as we dealt in Chap. 5 with the effect of noise upon the detection of amplitude modulation, we now take up its effect on the demodulation of frequency modulation. Here it is usual to pass the received signal through a limiter (Fig. 6-1) to remove all amplitude variations introduced by the noise, after filtering it to remove any noise outside the band occupied by the signal.

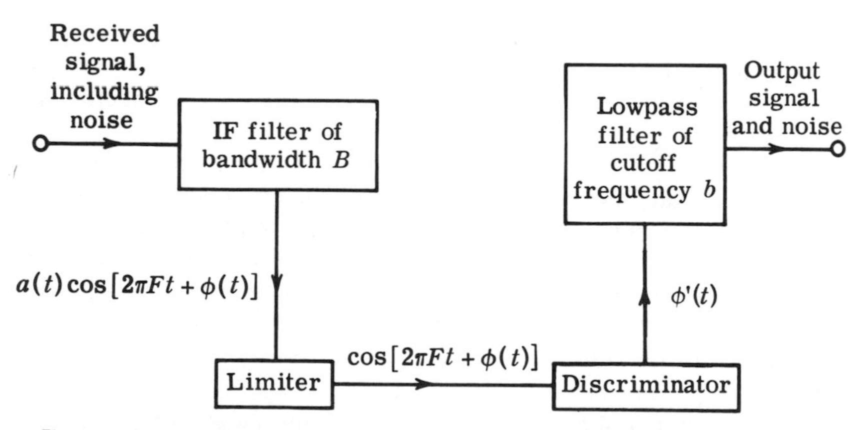


*Figure 6-1.* FM demodulation.

The constant-amplitude limiter output is fed to a discriminator, which consists of two resonant circuits, one tuned to each side of the input band, feeding opposing linear rectifiers, so that its output is proportional to the deviation of the instantaneous input frequency $F + \phi'(t)/2\pi$ from its average value $F$. Without loss of generality we may

† This chapter may be omitted on the first reading.

suppose the constant of proportionality is such that the discriminator output is simply the rate of change $\phi'(t)$ of the phase of the received signal (including the noise that passes through the intermediate-frequency filter).

Practical limiters and discriminators are unable to attain these ideals of perfectly constant output amplitude and perfect linearity, but they come sufficiently close that the mean-squared discriminator output, although not infinite, is extremely large on account of the very wide discriminator output spectrum. A low-pass output filter with cutoff frequency, say, $b$ is therefore used to remove output noise at frequencies above those contained in the desired output signal. Consequently, the limiter and discriminator might as well be ideal, and we shall assume them to be so in order to simplify our calculations.

## WIDE - BAND - FM QUIETING

For reasons which will become clear below, the frequency swing, and hence the bandwidth $B$, of the transmitted signal is often made large compared to the bandwidth $b$ of the modulating wave. This case is known as "large-index" or "wide-band" frequency modulation, although the transmitted signal is narrow-band, with $B \ll F$. For $b$ sufficiently small compared to $B$ we can calculate the output-noise power by multiplying the output bandwidth $2b$ (from $-b$ to $b$) by the output-power spectral density at zero frequency, $\Psi_{\phi'}(0)$, which is given by (4-54), together with (4-53) or (4-55).

Thus, when the input is a signal of amplitude $A$ and frequency $F$, together with gaussian noise having correlation function $\psi(\tau) \cos 2\pi F\tau$, the output-noise power is $2\pi^2 bV(A)$, which, as (4-53) shows, decreases with increasing $A$ and falls rapidly as $A$ becomes large compared to the rms input noise $\sigma$. In the absence of a signal this output-noise power is $2\pi^2 bV(0)$, with $V(0)$ given by (4-49) or the table. The quieting effect of the signal can therefore be described by the ratio of these two values of output noise power, namely, $V(0)/V(A)$.

**Strong Input Signal.** If we express the phase $\phi(t)$ of the received signal (Fig. 4-1) as the sum of two parts,

$$\phi(t) = \phi_1(t) + \phi_2(t),$$

where $\phi_2(t)$ is the multiple of $2\pi$ closest to $\phi(t)$ and $\phi_1(t)$ is the remainder, with absolute value no greater than $\pi$, that is, the principal value of (4-57), we find that $\mathrm{E}\,\{\phi_1{}^2\} \leq \pi^2$. Hence the integral of its power spectral density $\int_{-\infty}^{\infty} \Psi_{\phi_1}(f)\, df$ is no more than $\pi^2$, and $\Psi_{\phi_1}(f)$ must be of a smaller order than $1/f$ in the neighborhood of $f = 0$. Since, by (3-20), the power spectral density of $\phi_1'(t)$ is $\Psi_{\phi_1'}(f) = 4\pi^2f^2\Psi_{\phi_1}(f)$, it follows that $\Psi_{\phi_1'}(f)$ is of a smaller order than $f$ in the neighborhood of $f = 0$. Thus $\Psi_{\phi'}(0)$ is due entirely to the jumps of $\phi_2(t)$.

Referring to the derivation of (4-58), we see that these jumps of $\phi_2(t)$ represent rapid changes of $\phi(t)$ by approximately $\pm 2\pi$ when $A \gg \sigma$, that is, impulses of area approximately $\pm 2\pi$ in the discriminator output $\phi'(t)$. These are heard as distinct clicks when $A$ is sufficiently large, but as $A$ grows smaller they occur more frequently and merge first into a crackling noise and then into a continuous hiss.[1]

For sufficiently large $A$ the output-noise power due to the jumps of $\phi_2(t)$ thus becomes so small that we cannot ignore the contribution of $\phi_1(t)$, even though its power spectral density vanishes at $f = 0$. From Fig. 4-1 we see that, when $A$ is large, $\phi_1(t)$ is nearly always well approximated by $y(t)/A$, and hence has power spectral density $\Psi(f)/A^2$ (see p. 67). The power spectral density of its derivative is therefore

$$\Psi_{\phi_1'}(f) = \frac{4\pi^2f^2}{A^2}\Psi(f) \tag{6-1}$$

for $A \gg \sigma$. For $b \ll B$, this is well approximated in the passband of the output filter by $4\pi^2f^2\Psi(0)/A^2$, which is easily integrated with respect to $f$ from $-b$ to $b$ to give $8\pi^2b^3\Psi(0)/3A^2$ as the total output-noise power due to the fluctuations of $\phi_1(t)$.

A knowledge of the behavior of $\phi_2(t)$ tells us something about the behavior of $\phi_1(t)$ only in the immediate neighborhood of its jumps. Since these are rare when $A$ is large, we may regard $\phi_1'(t)$ and $\phi_2'(t)$ as statistically independent and simply add their contributions to the output-noise power at all frequencies of interest. The total output-noise power is thus better approximated by

$$2\pi^2bV(A) + \frac{8\pi^2b^3}{3A^2}\Psi(0) \tag{6-2}$$

for large $A$ than by $2\pi^2bV(A)$ alone, and the quieting effect of the signal on the output noise is therefore better expressed as

$$\frac{V(0)}{V(A) + (4b^2/3A^2)\Psi(0)}, \tag{6-3}$$

where $V(0)$ and $V(A)$ are (4-49) and (4-55), respectively, and $\Psi(0)$ is twice the input-noise power spectral density at the frequencies $\pm F$. Note

[1] S. O. Rice, Noise in FM Receivers, chap. 25, pp. 395–422, in M. Rosenblatt (ed.), "Time Series Analysis," John Wiley & Sons, Inc., New York, 1963. See also Gérard Battail, Sur le Seuil de Réception en Modulation de Fréquence, *Annales des Télécommunications*, **19**(1–2):21–48 (1964), who calls the $\phi_1'$ and $\phi_2'$ noise "bruit de voisinage" (neighborhood noise) and "bruit d'erreur," respectively.

that the second term in the denominator of (6-3) accurately describes the contribution of $\phi_1'(t)$ only for $A \gg \sigma$; for smaller $A$ this contribution is negligible compared to the first term, and the second term should be omitted. In this case $V(A)$ must be found by integrating (4-53) rather than by using the approximation (4-55).

## ☐ OUTPUT SIGNAL - TO - NOISE RATIO

Just as (5-4) is the gross low-frequency output signal in the case of amplitude demodulation, the signal output of our discriminator is

$$h = \mathrm{E}\,\{\phi'|D\} - \mathrm{E}\,\{\phi'\},$$

$\mathrm{E}\,\{\phi'|D\}$ being the average of $\phi'$ over the noise statistics for a specified $D$, and $\mathrm{E}\,\{\phi'\}$ being the average over both signal and noise statistics, where $D(t)$ is the frequency deviation of the input signal; that is, $F + D(t)$ is its instantaneous frequency. From (4-43) we see that this is

$$h = 2\pi(1 - e^{-A^2/2\sigma^2})(D - \mathrm{E}\,\{D\})$$

or simply

$$h = 2\pi(1 - e^{-A^2/2\sigma^2})D(t) \tag{6-4}$$

if, as we shall suppose, the average deviation is zero. Thus the output signal of an ideal discriminator is exactly proportional to $D(t)$, and there is no distortion. The output signal power is

$$\mathrm{E}\,\{h^2\} = 4\pi^2(1 - e^{-A^2/2\sigma^2})^2 D_{\mathrm{rms}}{}^2, \tag{6-5}$$

where $D_{\mathrm{rms}}$ is the root-mean-squared value of $D(t)$.

If this deviation is small enough not to affect the output-noise power significantly, the output signal-to-noise ratio is the quotient of (6-5) divided by (6-2),

$$R = \frac{2(1 - e^{-A^2/2\sigma^2})^2 D_{\mathrm{rms}}{}^2}{bV(A) + (4b^3/3A^2)\Psi(0)}. \tag{6-6}$$

As before, the second term of the denominator should be omitted when $A/\sigma$ is not large. This expression, which is valid only for wide-band FM, is shown graphically in Fig. 6-2$a$ for several values of $B/b$ as a function of the input signal-to-noise ratio $r = A^2/2\sigma^2$, both $r$ and $R$ being expressed in decibels. $D_{\mathrm{rms}}$ is assumed to be proportional to the intermediate-frequency bandwidth $B$ in this plot; ideally, $B$ should be just sufficient to accommodate the signal, whose spectral width is proportional to $D_{\mathrm{rms}}$ in the wide-band case, as (3-48) and (3-57) show.

For sufficiently large $r$, the first term of the denominator of (6-6) can be neglected, and we have

$$R = \frac{3\sigma^2 D_{\mathrm{rms}}{}^2}{b^3\Psi(0)}\,r; \tag{6-7}$$

that is, as the input-signal amplitude varies, $R$ varies directly as $r$, just as in the amplitude-demodulation case with large input signal-to-noise ratio. As $r$ falls, the ratio of the first term of the denominator of (6-6) to the second grows rapidly, and around a value of $r$ known as the *threshold* the first term very quickly becomes dominant. Thus just below the threshold $R$ drops sharply. As $r$ becomes small, however, the output-

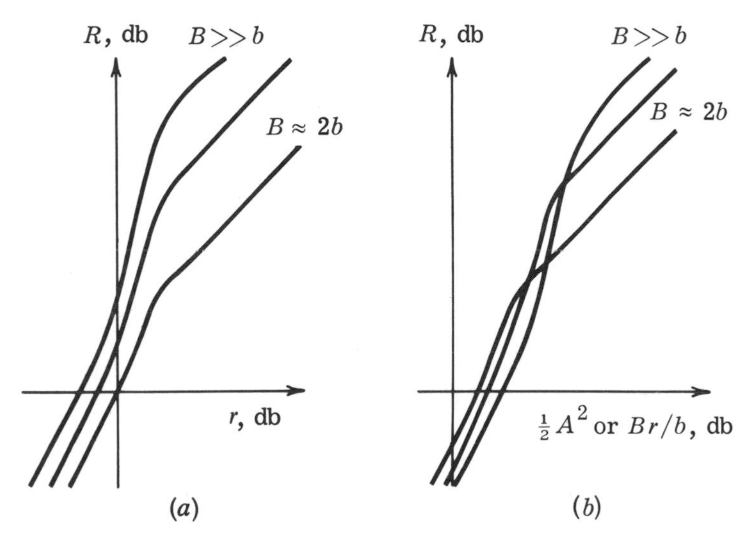


*Figure 6-2.* (*a*) Output signal-to-noise ratio versus input signal-to-noise ratio for various FM-signal bandwidths. (*b*) Output signal-to-noise ratio versus input signal power for various input bandwidths with a fixed input-noise power spectral density.

noise power stops increasing and the denominator of (6-6) approaches $bV(0)$, while its numerator becomes $2r^2D_{\text{rms}}{}^2$; that is,

$$R = \frac{2D_{\text{rms}}{}^2}{bV(0)}\, r^2, \tag{6-8}$$

which gives the curves of Fig. 6-2*a* a slope equal to 2 for small $r$, just as in amplitude demodulation.

These curves suggest that, when $r$ is small, $R$ is greater the larger is $B/b$. Although this conclusion is correct for fixed $B$, it is more usual to fix $b$ and to vary $B$, and, in doing so, we must take into account the fact that $r$ will vary inversely as $B$ for a fixed input-noise power spectral density. To take this into account, $R$ has been plotted in Fig. 6-2*b* as a function of the input signal power $\frac{1}{2}A^2$, which is proportional to $Br/b$. The effect is to shift each curve of Fig. 6-2*a* to the right by $10 \log_{10} B/b$ decibels, showing that, for small $r$, the greater $B/b$ is, the smaller $R$ is.

**Problem 6-1.** Show that, if the discriminator output filter's frequency response $|G(f)|^2$ is not flat from 0 to $b$ and zero for $f > b$ (that is, is not an ideal low-pass response), the first $b$ in the denominator of (6-6) should be replaced by

$$b_2 = \int_0^\infty |G(f)|^2 \, df/|G(0+)|^2$$

and the second by

$$b_1 = \sqrt[3]{3 \int_0^\infty f^2 |G(f)|^2 \, df/|G(0+)|^2},$$

in which $|G(0+)|^2$ indicates the filter's response to the frequencies of the signal's modulating waveform rather than to direct current. Both of these effective output bandwidths must be small compared to the limiter's input bandwidth for (6-6) to be applicable. Notice the importance of the way $|G(f)|^2$ falls off at high frequencies; with a simple $RC$ low-pass filter, for example, $b_1$ is infinite.

**Problem 6-2.** Plot (6-6) in the neighborhood of the threshold for the fifth input-noise spectrum of the table on page 71, and notice that the threshold effect becomes more pronounced as $B/b$ increases.

**Problem 6-3.** Defining the FM threshold as the (larger) value of $r$ for which the two terms of the denominator of (6-6) are equal, show that it satisfies the equation $\sqrt{r}\, e^{-r} = \sqrt{\pi}\, b^2\Psi(0)/6\rho\sigma^2$ and evaluate it for the case of the rectangular input-noise spectrum with $B/b = 15$.

**Problem 6-4.** Making use of the derivation of (4-58), determine the average rate of output noise clicks at the threshold found in Prob. 6-3 with $B = 225$ kc. Notice that they are too frequent to be heard as distinct clicks but are infrequent enough to validate the assumption, used in deriving (4-58), that they form a Poisson process. What will be the rate of clicks 3 db above and 3 db below the threshold?

**Problem 6-5.** Show that, for $r$ above the threshold, the quieting (6-3) has the form $k\rho^2 r/b^2$, and determine the constant $k$ for each of the six input-noise spectra of the table on page 71.

**The Effect of Input-Signal Modulation on the Output Noise.** When the input signal-to-noise ratio $r = A^2/2\sigma^2$ is small, the input signal has little effect on the output noise, and consequently (6-6) (without the second term of its denominator) is valid for *any* $D_{\text{rms}}$, provided only that the i-f bandwidth $B$ is sufficient to pass the signal. When $r$ is not small, however, the result of Prob. 4-20 must be used to take account of the effect of input-signal modulation on the output noise power if $D_{\text{rms}}$ is not small compared to $B$. This result should be averaged over the distribution of $D$ for the same reason that (3-58) is an average over $A$ and $D$; the assumption that $D$ varies slowly, which underlies this averaging, is justified by the large value of $B/b$.

The $\Psi_{\phi'}(0+)$ of Prob. 4-20 (the + avoids the second term of the spectrum of Prob. 3-13, which represents the d-c output component) is valid when $r$ is large, even though it may not exceed the threshold. In this case, with $D$ comparable to $\rho$, as is usual when the i-f bandwidth $B \gg b$ matches the spectral width of the signal, the first term of $\Psi_{\phi'}(0+)$ is small compared to the second, which gives us

$$\Psi_{\phi'}(0+) = 4\pi^2 e^{-r}|D|.$$

Averaging and multiplying by the output bandwidth $2b$, we obtain the approximation

$$8\pi^2 b e^{-r} \, \mathrm{E}\,\{|D|\} \tag{6-9}$$

for the output noise power due to jumps of $\phi_2(t)$.

To find the $\phi_1'(t)$ output noise when the input signal is modulated, we suppose that its frequency is $F + D$. With this as reference frequency and a suitable choice of reference phase, the signal can be represented by a real phasor $A$. The noise, which is represented by the phasor $x(t) + iy(t)$ in terms of the center of symmetry of its spectrum, $F$, will be represented by $x_D(t) + iy_D(t)$, with

$$\begin{aligned} x_D(t) &= x(t) \cos 2\pi Dt + y(t) \sin 2\pi Dt, \\ y_D(t) &= -x(t) \sin 2\pi Dt + y(t) \cos 2\pi Dt, \end{aligned} \tag{6-10}$$

since $(x + iy)e^{2\pi iFt} = (x_D + iy_D)e^{2\pi i(F+D)t}$. A phasor diagram like Fig. 4-1 shows that, for $A \gg \sigma$, the phase angle of the total differs from that of the signal by approximately

$$\phi_1(t) = \frac{y_D(t)}{A}$$

plus a multiple $\phi_2(t)$ of $2\pi$. Since the signal has phase $2\pi Dt$ in terms of reference frequency $F$, the phase of the sum of signal and noise is

$$\phi(t) = 2\pi Dt + \phi_1(t) + \phi_2(t).$$

The last term accounts for the foregoing output noise, (6-9), and, along with the term $2\pi Dt$, yields the output signal (6-4) (see Prob. 4-20). The term $\phi_1(t)$ gives output noise $y_D'(t)/A$, whose power spectral density is $4\pi^2 f^2/A^2$ times that of $y_D(t)$. Since $x(t)$ and $y(t)$ are statistically independent, each with correlation function $\psi(\tau)$, it follows from (6-10) that $y_D(t)$ has correlation function $\psi(\tau) \cos 2\pi D\tau$ and hence power spectral density $\Psi_{y_D}(f) = \frac{1}{2}\Psi(f - D) + \frac{1}{2}\Psi(f + D)$. The power spectral density of the discriminator output noise due to $\phi_1(t)$ is therefore

$$\Psi_{\phi_1'}(f) = \frac{2\pi^2 f^2}{A^2} [\Psi(f - D) + \Psi(f + D)], \tag{6-11}$$

which is shown in Fig. 6-3 for the case of a rectangular input-noise spectrum.

For this case we have

$$\Psi(f) = \begin{cases} \sigma^2/B & \text{for } |f| < \frac{1}{2}B, \\ 0 & \text{otherwise.} \end{cases} \tag{6-12}$$

Provided that

$$b \leq \tfrac{1}{2}B - |D|, \tag{6-13}$$

we can readily integrate (6-11) to obtain the resulting output-noise power,

$$\int_{-b}^{b} \Psi_{\phi_1'}(f)\, df = \frac{4\pi^2 b^3}{3Br}, \tag{6-14}$$

in this case. Since (6-14) is independent of $D$, it needs no averaging, the condition (6-13) being satisfied for wideband FM.

The output signal-to-noise ratio is therefore the ratio of (6-5), which is approximately $4\pi^2 D_{\text{rms}}{}^2$ for large $r$, to the sum of (6-9) and (6-14), namely,

$$R = \frac{3BD_{\text{rms}}{}^2 r}{6Bbre^{-r}\,\mathrm{E}\,\{|D|\} + b^3}. \tag{6-15}$$

Because of the modulation, the threshold is here somewhat higher than in the case of (6-6). Below it the second term of the denominator can (and should) be neglected. Above the threshold (6-15) becomes

$$R = \frac{3BD_{\text{rms}}{}^2 r}{b^3}, \tag{6-16}$$

which is considerably greater than $r$ when $B$ and $D_{\text{rms}}$ are large compared to $b$. In fact, (6-16) is $3D_{\text{rms}}{}^2/b^2$ times $Br/b$, the ratio of the input-signal power to the input-noise power in a band of width $b$—just sufficient for a single-sideband signal carrying the same modulating wave. (Single-sideband signals are demodulated by a downward shift in their spectrum, which does not change the signal-to-noise ratio.) This improvement by the factor $3D_{\text{rms}}{}^2/b^2$ in the case of wide-band FM when $r$ exceeds the threshold is attributable to the fact that (6-11), like (6-1), is very small near $f = 0$ and hence only a very small fraction of the $\phi_1'(t)$ noise is passed by the output filter.

We have found the output signal-to-noise ratio of a wide-band FM receiver for all values of $r$ when $D_{\text{rms}} \ll B$ and for small $r$ and large $r$—whether below, at, or above the threshold—when $D_{\text{rms}}$ is not small. In

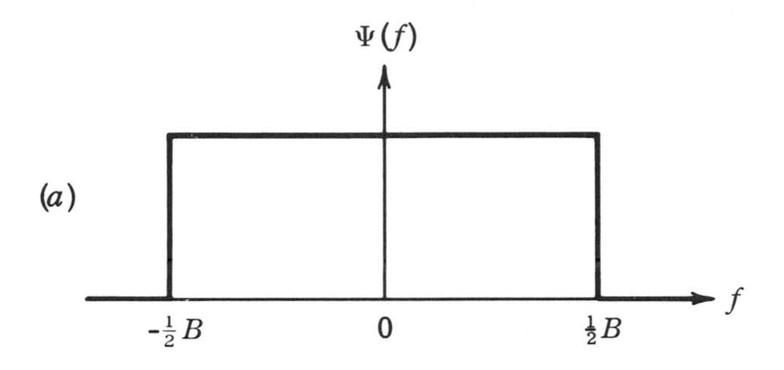


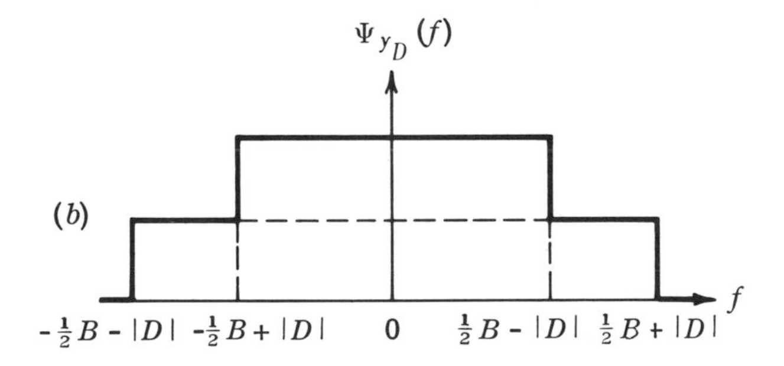


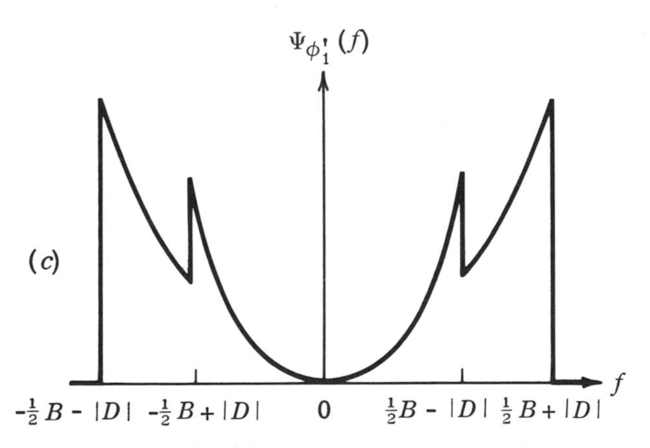


**Figure 6-3.** Spectra resulting from a constant frequency deviation $D$ and a rectangular input-noise spectrum of width $B$: (a) $\Psi(f)$, the power spectral density of $x(t)$ and $y(t)$, (b) $\Psi_{y_D}(f) = \frac{1}{2}\Psi(f - D) + \frac{1}{2}\Psi(f + D)$, the power spectral density of $x_D(t)$ and of $y_D(t)$, and (c) $\Psi_{\phi_1'}(f)$, the discriminator output-noise spectrum resulting from $\phi_1(t) \cong y_D(t)/A$ for large signal amplitude $A$.

*any* case, (6-5) correctly gives the output signal power, but when $A$ is comparable with $\sigma$ and $D_{\text{rms}}$ is not small, the output-noise power is an unsolved problem. When $r$ exceeds the threshold, (6-11) can be used to find the output-noise power, even in the narrow-band case, but for smaller $r$ the output-noise power is difficult to determine when $B/b$ is not large, even when $r$ is zero, since it requires numerical Fourier transformation of the correlation function.[1]

Still more difficult is the case of imperfect limiting and a nonlinear discriminator characteristic. It is clear, however, that discriminator characteristics that droop for very large frequency deviations, as those of practical discriminators do, will reduce the impulsive output noise by effectively limiting the amplitude of the clicks. Imperfect limiting will likewise reduce the $\phi_2'(t)$ output noise by delivering to the discriminator a smaller input amplitude at the times when $\phi_2(t)$ is likely to jump by $\pm 2\pi$. In addition, a nonlinear discriminator characteristic will result in distortion of the output signal—at least when the signal-to-noise ratio is not high.

**Problem 6-6.** Calculate the value of $r$ for which the two terms of the denominator of (6-15) are equal when $B = 15b$ and $D$ is sinusoidal with amplitude $5b$. Notice that it exceeds the threshold for (6-6), found in Prob. 6-3, by 0.9 db.

**Problem 6-7.** Show that, if the $\phi_2'(t)$ noise can be neglected, the total (unfiltered) discriminator output-noise power is $2\pi^2(\rho^2 + D_{\text{rms}}{}^2)/r$, and hence its output signal-to-noise ratio is $2rD_{\text{rms}}{}^2/(\rho^2 + D_{\text{rms}}{}^2)$, which exhibits no signal-to-noise-ratio improvement.

**Problem 6-8.** From (6-5), (6-11), and (6-12) find the output signal-to-noise ratio above the threshold when $D$ varies sinusoidally with amplitude greater than $\frac{1}{2}B - b$.

## ☐ OMITTING THE LIMITER

If we omit the limiter in Fig. 6-1 and feed the sum $a(t) \cos [2\pi Ft + \phi(t)]$ of input signal and noise directly to the discriminator, its output will be proportional not only to the instantaneous frequency deviation $\phi'(t)/2\pi$ of its input as before but also to the amplitude $a(t)$ of its input. Without loss of generality, we can thus suppose its output to be $a\phi'$, which, we see

[1] S. O. Rice, Statistical Properties of a Sine Wave Plus Random Noise, *Bell System Tech. J.*, **27**:109–157 (1948); M. C. Wang in J. L. Lawson and G. E. Uhlenbeck (eds.), "Threshold Signals," MIT Radiation Laboratory Series, vol. 24, pp. 369–383, McGraw-Hill Book Company, New York, 1950.

from Fig. 6-4, is the tangential component (in the direction $\phi + \frac{1}{2}\pi$) of the rate of change of the phasor $ae^{i\phi}$ representing its input.[1]

If the frequency of the input signal is $F + D$, it can be represented by the phasor $Ae^{2\pi iDt}$, which rotates counterclockwise with angular velocity $2\pi D$. Its linear velocity is thus $2\pi AD$ in the direction perpendicular to itself. If $\theta$ is the angle between the signal phasor and the total-input phasor (Fig. 6-4), this velocity contributes $2\pi AD \cos\theta$ to $a\phi'$. Similarly, the rate of change of the noise phasor $x(t) + iy(t)$ contributes

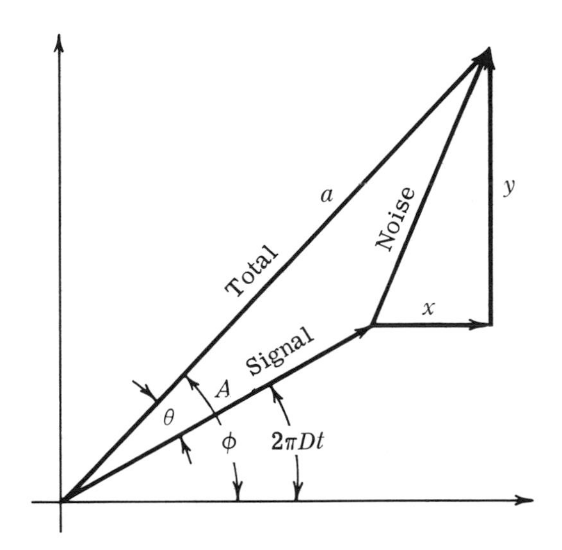


*Figure 6-4.* Phasor representation of a signal of frequency $F + D$ and narrow-band noise with spectrum symmetric about the frequency $F$.

$-x'(t)\sin\phi(t) + y'(t)\cos\phi(t)$, and the total discriminator output is therefore

$$a\phi' = 2\pi AD\cos\theta(t) - x'(t)\sin\phi(t) + y'(t)\cos\phi(t). \qquad (6\text{-}17)$$

**Output Signal.** As in the case of (6-4), the output signal is the mean discriminator output, $\mathrm{E}\,\{a\phi'|D\}$. Since $x'(t)$ and $y'(t)$ have mean value zero and are statistically independent of $x(t)$ and $y(t)$ [which determine $\phi(t)$ and $\theta(t)$] at the same instant, the mean value of the last two terms of (6-17) is zero. Because $x(t)$ and $y(t)$ obey the same circular normal distribution, regardless of the reference phase, the distribution

[1] N. M. Blachman, The Demodulation of an FM Carrier and Random Noise by a Limiter and Discriminator, *J. Appl. Phys.*, **20**:38–47 (1949), and The Demodulation of a Frequency-Modulated Carrier and Random Noise by a Discriminator, *J. Appl. Phys.*, **20**:976–983 (1949).

of $\theta(t)$ does not depend on the phase $2\pi Dt$ of the input signal. For convenience we therefore suppose this phase to be zero so that $\theta = \phi$ and Fig. 6-4 looks like Fig. 4-1.

Making use of the value of E $\{\cos \phi\}$ found in Prob. 4-6, we thus see that the discriminator's output signal is

$$\mathrm{E}\ \{a\phi'|D\} = \pi AD \sqrt{\pi r}\ e^{-r/2}[I_0(\tfrac{1}{2}r) + I_1(\tfrac{1}{2}r)]. \tag{6-18}$$

For our idealized discriminator, (6-17) and hence (6-18) are valid, whether the input-signal deviation $D$ is constant or is varying. Here again we see that the signal is demodulated without distortion, since (6-18) is proportional to $D$. As before, we are supposing that the mean value of $D$ is zero, and so (6-18) contains no d-c component. Thus, averaging over the distribution of $D$, we find that the output-signal power is

$$\pi^3 A^2 D_{\mathrm{rms}}{}^2 r e^{-r}[I_0(\tfrac{1}{2}r) + I_1(\tfrac{1}{2}r)]^2. \tag{6-19}$$

**Output Noise.** The total discriminator output power is the mean-squared value of (6-17). Averaging first with respect to $x'$ and $y'$, which are independent of each other and of the other variables and have mean-squared value (4-17) $4\pi^2\rho^2\sigma^2$, we find that

$$\mathrm{E}\ \{a^2\phi'^2\} = 4\pi^2A^2D^2\ \mathrm{E}\ \{\cos^2\theta\} + 4\pi^2\rho^2\sigma^2.$$

Again putting $\theta = \phi$ and making use of the value of E $\{\cos^2 \phi\}$ found in Prob. 4-6, we find that the total discriminator output power is

$$\mathrm{E}\ \{a^2\phi'^2\} = 4\pi^2A^2D^2\left(1 - \frac{1 - e^{-r}}{2r}\right) + 4\pi^2\rho^2\sigma^2. \tag{6-20}$$

Because the output signal and output noise are orthogonal, we obtain the total output-noise power by subtracting (6-19) from (6-20) and averaging over the distribution of $D$. Dividing (6-19) by the result, we get the ratio of output-signal power to total output-noise power,

$$R = \frac{\tfrac{1}{2}\pi r^2 e^{-r}[I_0(\tfrac{1}{2}r) + I_1(\tfrac{1}{2}r)]^2}{(\rho/D_{\mathrm{rms}})^2 + e^{-r} + 2r - 1 - \tfrac{1}{2}\pi r^2 e^{-r}[I_0(\tfrac{1}{2}r) + I_1(\tfrac{1}{2}r)]^2}. \tag{6-21}$$

**Large Input Signal-to-Noise Ratio.** The effect of the output filter (Fig. 6-1) is not included in (6-21). To include it we need to know the discriminator's output-noise spectrum, which in general is difficult to determine but can readily be found when the input signal-to-noise ratio is high. In this case, the $\cos\theta$ in (6-17) is very close to unity, and the $\phi$ in its last two terms can be replaced by $2\pi Dt$ (Fig. 6-4). Thus the discriminator output is

$$a\phi' \cong 2\pi AD - x'(t) \sin 2\pi Dt + y'(t) \cos 2\pi Dt. \tag{6-22}$$

The first term on the right-hand side is the output signal, in agreement with (6-18) in view of (4-28). The last two terms represent the output noise, with total power $4\pi^2\rho^2\sigma^2$ independent of the signal's modulation. Making use of the fact that $x'(t)$ and $y'(t)$ are independent, zero-mean random processes, each with correlation function $-\psi''(\tau)$ by (3-19), we find that this output noise has correlation function

$$\psi_n(\tau) = -\psi''(\tau) \cos 2\pi D\tau.$$

By the Wiener-Khinchin theorem, then, its power spectral density is

$$\Psi_n(f) = 2\pi^2(f - D)^2\Psi(f - D) + 2\pi^2(f + D)^2\Psi(f + D), \quad (6\text{-}23)$$

which is the result of shifting half the power spectral density of $x'$ (or of $y'$) $D$ to the right and half $D$ to the left. This is shown in Fig. 6-5 for the case of a rectangular input-noise spectrum, (6-12).

Unlike (6-11), (6-23) does not vanish for $f = 0$, and hence the wide-band FM signal-to-noise-ratio improvement described by (6-16) cannot be obtained when the limiter is omitted. In fact, for the rectangular input-noise spectrum (6-12), (6-23) is $4\pi^2D^2\sigma^2/B$ for $f = 0$, which gives low-pass-filter output-noise power $8\pi^2 D_{\text{rms}}{}^2\sigma^2 b/B$ if $b \ll B$ and $D$ can consequently be regarded as varying slowly.

Dividing the output-signal power, $4\pi^2A^2D_{\text{rms}}{}^2$, by this output-noise power, we get the output signal-to-noise ratio

$$R = Br/b, \quad (6\text{-}24)$$

which simply shows the improvement to be expected when reducing the bandwidth from $B$ to $b$. There is no signal-to-noise-ratio improvement by the factor $3D_{\text{rms}}{}^2/b^2$ and, consequently, no threshold effect without the limiter. Here also, jumps in the value of $\phi_2$ (better described as jumps in the nearest multiple of $2\pi$ to $\theta$ when the input signal is modulated) are less important than when a limiter is included, for with $r$ large these jumps are nearly always accompanied by a very small value of $a$, which reduces the discriminator output and precludes any clicking.

In the narrow-band case, i.e., with $B$ comparable to $2b$, the deviation $D$ cannot be regarded as changing only slowly. However, the average of (6-23) over its distribution remains a good approximation to the output-noise spectrum since, as we have seen, the total output-noise power is not affected by the (spectral shape of the) modulation, and its integral over the pass band of the output filter will approximate well the output-noise power. [For a similar reason (see Prob. 6-7), the average of (6-11) over the distribution of $D$ remains a good approximation to the output-noise spectrum even in the narrow-band case when $r$ exceeds the threshold and a limiter is used.] In *any* case, the output-signal power is given by (6-19).

The investigation of the output noise of an FM receiver and its spectrum in the general case, with or without a limiter, involves con-

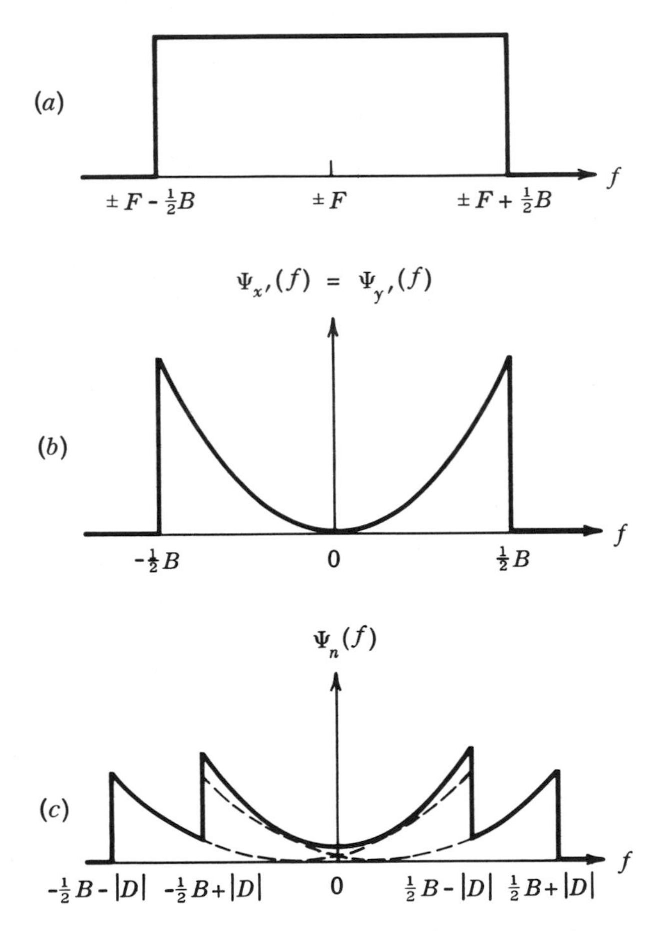


***Figure 6-5.*** When the input to a limiterless discriminator is a strong signal of constant deviation $D$ plus narrow-band noise having a rectangular spectrum $(a)$ of width $B$, the resulting spectrum of $x'(t)$ and of $y'(t)$ is $(b)$, and the discriminator output-noise spectrum is the solid curve of $(c)$, which is the sum of the two dashed curves—each a shifted version of $(b)$.

siderably more difficult calculations than those of the limiting cases we have treated, but Middleton[1] has obtained a number of useful though complicated results. The reception of a wide-band FM signal by a receiver whose local oscillator is made to track the signal's frequency by means of

[1] Middleton, *op. cit.*, chap. 15.

feedback from the discriminator output can easily be treated by applying the results of this chapter if the output signal-to-noise ratio is high. However, the theoretical determination of its performance under other circumstances remains an unsolved problem.

**Problem 6-9.** By integrating (6-23) from $f = -b$ to $f = b \leq \frac{1}{2}B - |D|$ for the case of (6-12), obtain a better approximation to the output signal-to-noise ratio than (6-24), namely, $R = (3Br/b)D_{\text{rms}}^2/(3D_{\text{rms}}^2 + b^2)$, which is always worse than (6-16). Note, however, that the ratio of the output-signal power to the discriminator's total output-noise power is $R = 2rD_{\text{rms}}^2/\rho^2$, which is always better than the value found in Prob. 6-7 with the limiter included.

**Problem 6-10.** Find the discriminator's output-signal power and total output-noise power when a power-law limiter (5-7) (followed by a fundamental-zone filter) is used in place of the ideal ($n = 0$) limiter.

**Problem 6-11.** Find the average of (6-23) with (6-12) over the distribution of $D$ when the modulation is as in Prob. 6-8.

# chapter 7 SIGNAL DETECTION AND RECOGNITION

The output signal-to-noise ratio of a receiver, which we studied in Chapters 5 and 6, indicates the accuracy with which the receiver's output reproduces the instantaneous values of the modulating wave. Being only a single number, however, this ratio does not specify the spectrum or other statistics of the output noise and distortion which, considered in the light of what is done with the receiver's output, help to determine how troublesome the noise actually is. What is more important is a numerical measure of this troublesomeness, if a suitable measure can be found. Such a measure is readily available in the case of binary signaling, where the troublesomeness is simply a matter of incorrect decisions by the receiver as to which of the two signals was sent. Accordingly, we turn our attention in this chapter to binary signaling.

## ☐ BINARY SIGNALING WITH EXACTLY KNOWN SIGNALS

Binary signaling involves the transmission of one of two waveforms, say $s_1(t)$ and $s_0(t)$, which are often designated *mark* and *space*, respectively, and we begin by supposing that these two waveforms are known exactly to the receiver designer, including their time of occurrence. In the case of "on-off keying," the space signal $s_0(t)$ is identically zero, and the receiver must decide on the basis of the received signal, including noise, whether $s_1(t)$ was transmitted or not. With *phase-shift keying* (PSK) or *frequency-shift keying* (FSK), on the other hand, the two waveforms differ only in phase or in frequency, respectively.

In general, the received signal $u(t)$ is either $s_1(t)$ or $s_0(t)$ corrupted by noise, and the receiver must infer from the waveform $u(t)$ (for all $t$) which of the two was transmitted. That is, it must divide the set of all possible waveforms $u(t)$ into two subsets—those to be regarded as resulting from the transmission of $s_1(t)$ and those to be regarded as due to the transmission of $s_0(t)$.

**The Weight of Evidence.** If $P(1)$ and $P(0)$, with $P(1) + P(0) = 1$, are the prior probabilities that mark and space, respectively, were sent [i.e., prior to observing $u(t)$], the prior odds in favor of mark are $P(1):P(0)$. Denoting by $P(1|\mathbf{u})$ and $P(0|\mathbf{u})$ the respective posterior probabilities based on the entire curve of $u(t)$, with $P(1|\mathbf{u}) + P(0|\mathbf{u}) = 1$, we see that the posterior odds in favor of mark are $P(1|\mathbf{u}):P(0|\mathbf{u})$.

Representing by $p(\mathbf{u}|1)$ and $p(\mathbf{u}|0)$ the probability densities (in an appropriate sense)[1] of $u(t)$ when mark and space, respectively, are sent, and making use of Bayes's rule, which states that

$$P(k|\mathbf{u})p(\mathbf{u}) = p(\mathbf{u}|k)P(k)$$

for $k = 0$ or 1, we see that the posterior odds are related to the prior odds by

$$\frac{P(1|\mathbf{u})}{P(0|\mathbf{u})} = \frac{P(1)}{P(0)} \cdot \frac{p(\mathbf{u}|1)}{p(\mathbf{u}|0)}. \tag{7-1}$$

Quantities like $p(\mathbf{u}|1)$ and $p(\mathbf{u}|0)$ are called *likelihoods*, and their ratio is known as the *likelihood ratio*. Taking logarithms, we get

$$\ln \frac{P(1|\mathbf{u})}{P(0|\mathbf{u})} = \ln \frac{P(1)}{P(0)} + \ln \frac{p(\mathbf{u}|1)}{p(\mathbf{u}|0)}. \tag{7-2}$$

The left-hand side of (7-2) is described[2] as the posterior or total "weight of evidence" in favor of mark. The first term on the right is the prior weight of evidence, and the last term is the weight of evidence provided by $\mathbf{u}$. This last term thus summarizes *all* of the information that $u(t)$ gives concerning which signal was sent; from it and the prior probabilities of mark and space we can find their posterior probabilities. This is as much as we can hope to find out about which of them was transmitted.

If the total weight of evidence is great enough, i.e., if the posterior odds are sufficiently favorable, the inference should be drawn that a mark was sent. The appropriate threshold for this decision will depend on the weights given to the two kinds of correct inferences and to the two kinds of errors. Thus a contour of the likelihood ratio serves to separate the set of all possible received signals $u(t)$ optimally into two subsets, one to be regarded as resulting from the transmission of mark and the other of space. To make full use of the information available to it from $u(t)$,

[1] If the received signal $u(t)$ can be adequately described by a finite number of parameters comprising a vector $\mathbf{u}$, then $p(\mathbf{u}|k)$ will be the joint probability density function of these parameters.

[2] I. J. Good, "Probability and the Weighing of Evidence," Hafner Publishing Company, Inc., New York, 1950.

a receiver must therefore evaluate the logarithm (or some other monotonic function) of the likelihood ratio and must compare it with a suitable threshold; such a receiver is called *optimum.*

**Problem 7-1.** Show that, if $C_{11}$ and $C_{00}$ are the credit gained for correct decisions, and $C_{10}$ and $C_{01}$ are the costs incurred for calling a mark a space and vice versa, respectively, the expected gain is maximized by basing the decision on whether the logarithm of the likelihood ratio exceeds

$$\ln \frac{C_{00} + C_{01}}{C_{10} + C_{11}} - \ln \frac{P(1)}{P(0)}.$$

(HINT: When we infer a mark from $u(t)$, our expected gain is $P(1|\mathbf{u})C_{11} - P(0)|\mathbf{u})C_{01}$.) Show that in general this threshold does not result in a proportion $P(1)$ of mark decisions. Notice that, if we do not know $P(1)/P(0)$, and we decide to attribute to it the least favorable value for each possible choice of threshold, the best resulting ("minimax") value for the threshold provides the best possible lower bound for the expected gain.

**Problem 7-2.**[1] Let $P_F$ and $P_M$ denote the probabilities that, with some given decision criterion, $s_0(t)$ corrupted by noise will be incorrectly interpreted as resulting from the transmission of $s_1(t)$ and vice versa, respectively. Taking the net cost when $s_0(t)$ is sent to be a nondecreasing function $C_0(P_F)$ and the net cost when $s_1(t)$ is sent to be a nondecreasing function $C_1(P_M)$, differentiate the expected net cost with respect to the location of the boundary between the two subsets of $u(t)$. Show that it is minimized by using a likelihood-ratio contour as the boundary. Note that this generalization of Prob. 7-1, which used a "Bayes criterion," covers the case where the "false-alarm probability" $P_F$ is specified and the "miss probability" $P_M$ is to be minimized ("Neyman-Pearson criterion") or vice versa. How can the case where $P_F + P_M$ is specified ("ideal observer") be included?

## ☐ THE SAMPLING THEOREM

We now derive the sampling theorem,[2] which asserts that, in the band-limited case, $u(t)$ is determined completely by a countable set of values and thus will enable us to express the likelihood ratio $p(\mathbf{u}|1)/p(\mathbf{u}|0)$. Postponing until later the case in which $u(t)$ has no Fourier transform, we suppose that its Fourier transform is $U(f)$, and we form

$$U_{2W}(f) = \sum_{j=-\infty}^{\infty} U(f - 2jW), \tag{7-3}$$

[1] Ivan Selin, "Detection Theory," pp. 8–15, Princeton University Press, Princeton, N.J., 1965; D. Middleton, "An Introduction to Statistical Communication Theory," chap. 19, McGraw-Hill Book Company, New York, 1960.

[2] C. E. Shannon, Communication in the Presence of Noise, *Proc. IRE,* **37**:10–21 (1949).

the summation of $U(f)$ displaced by all multiples of $2W$ for some $W > 0$. This quantity will therefore be a periodic function with period $2W$, which can be expressed as an exponential Fourier series. Because $\int_{-\infty}^{\infty} U(f)e^{2\pi ift}\,df = u(t)$, the Fourier coefficients $\int_{-W}^{W} U_{2W}(f)e^{k\pi if/W}\,df/2W$ turn out to be $u(k/2W)/2W$, and we have

$$U_{2W}(f) = \frac{1}{2W} \sum_{k=-\infty}^{\infty} u\left(\frac{k}{2W}\right) e^{-k\pi if/W}.$$

If $U(f) = 0$ for all $|f| \geq W$, then we see from (7-3) that $U_{2W}(f) = U(f)$ for all $|f| < W$, and so

$$u(t) = \int_{-W}^{W} U_{2W}(f)e^{2\pi ift}\,df.$$

In general, however, this integral will not give us $u(t)$, and we denote it instead by $v(t)$, the inverse Fourier transform of

$$V(f) = U_{2W}(f) \operatorname{rect} \frac{f}{2W} \tag{7-4}$$

with rect $(f/2W)$ equal to 1 for $-W < f < W$ and 0 otherwise, by (3-32). Calculating the inverse Fourier transform with $U_{2W}(f)$ replaced by its exponential Fourier expansion, we find

$$v(t) = \sum_{k=-\infty}^{\infty} u\left(\frac{k}{2W}\right) \operatorname{sinc}(2Wt - k), \tag{7-5}$$

where $\operatorname{sinc} x = (\sin \pi x)/\pi x$ is the Fourier transform (3-31) of rect $y$.

When $U(f) = 0$ for all $|f| \geq W$, it follows from (7-4) that $V(f)$ and $U(f)$ are identical and hence that $v(t) = u(t)$. Thus any finite-energy waveform $u(t)$ whose spectrum is confined to the band from $-W$ to $W$ can be expressed as

$$u(t) = \sum_{j=-\infty}^{\infty} u(j/2W) \operatorname{sinc}(2Wt - j), \tag{7-6}$$

which involves only those values of $u(t)$ with $t$ a multiple of $1/2W$. Such values are often called "samples" of $u(t)$, and (7-6) is known as the (or a) sampling theorem. It serves to interpolate between sample values, giving $u(t)$ for *any* $t$ in terms of them.

**Aliasing.** Regardless of the spectrum of $u(t)$, the spectrum (7-4) of $v(t)$ is confined to the band from $-W$ to $W$. If the spectrum of $u(t)$ is not confined to this band, therefore, $v(t)$ will differ from $u(t)$. In particular, as we see from (7-4), any component of $u(t)$ of frequency $f$ outside this band will contribute to $v(t)$ a component of frequency $f - 2jW$, where $j$ is the closest integer to $f/2W$. This translation of frequency components effected by (7-5) is called *aliasing*.

As (7-5) indicates, $v(t)$ can be obtained from $u(t)$ by sampling $u(t)$ every $1/2W$ seconds, forming a short pulse of area $u(t)$ at each sampling time, and passing the resulting pulse train through an ideal low-pass filter with cutoff frequency $W$. Its output will be $u(t)$ (delayed by the propagation time of the filter) if $u(t)$ contains no frequency higher than $W$; otherwise, aliasing will occur.

When $u(t)$ is an ergodic random process, for example, as we saw in Chap. 3, it has no Fourier transform $U(f)$. In this case, we deal instead with the truncated $u_T(t)$, (3-4), which has Fourier transform $U_T(f)$. We assume that the power spectral density $\Psi_u(f)$ vanishes for $|f| \geq W$, but the truncation of $u(t)$ nevertheless spreads the spectrum $U_T(f)$ beyond $W$ and results in aliasing. However, by (3-10), as $T$ grows infinite, the expected proportion of the total energy that is aliased goes to zero, and we conclude that in the limit the mean-squared difference between the two sides of (7-6) is zero.

Since (7-6) is thus valid for any band-limited ergodic random process, as well as for any $u(t)$ having a band-limited Fourier transform $U(f)$, it is also applicable to the sum of two such waveforms.

**Problem 7-3.** By applying the sampling theorem to the correlation function of a random process $u(t)$ whose power spectral density vanishes at all frequencies above $W$, show that the mean-squared difference between $u(t)$ and $\sum_{j=-\infty}^{\infty} u(j/2W) \operatorname{sinc}(2Wt - j)$ is zero.[1]

**Energy and Cross-Correlation.** Because $\operatorname{rect}^2 f = \operatorname{rect} f$, we have $\operatorname{sinc} t \star \operatorname{sinc} t = \operatorname{sinc} t$ and, more specifically, for any integers $j$ and $k$,

$$\int_{-\infty}^{\infty} \operatorname{sinc}(t - j) \operatorname{sinc}(t - k)\, dt = \begin{cases} 1 & \text{with } j = k, \\ 0 & \text{with } j \neq k. \end{cases} \tag{7-7}$$

Using this "orthonormal" property of the sinc function, we find that the total energy of the waveform (7-6) is, say,

$$\begin{aligned} E_u &= \int_{-\infty}^{\infty} u^2(t)\, dt \\ &= \frac{1}{2W} \sum_{j=-\infty}^{\infty} u^2(j/2W) \\ &= \tilde{\mathbf{u}}\mathbf{u}, \end{aligned} \tag{7-8}$$

[1] L. A. Wainstein and V. D. Zubakov, "Extraction of Signals from Noise," chap. 3, sec. 19, translated by R. A. Silverman, Prentice-Hall, Inc., Englewood Cliffs, N.J., 1962.

where **u** is a vector (single-column matrix) with components

$$\frac{u(j/2W)}{\sqrt{2W}}, \qquad j = \ldots, -2, -1, 0, 1, 2, \ldots .$$

Similarly, if $s(t)$ contains no frequency higher than $W$, then

$$s(t) = \sum_{k=-\infty}^{\infty} s(k/2W) \operatorname{sinc}(2Wt - k),$$

and we have for the cross-correlation (with displacement $\tau = 0$), say,

$$\begin{aligned} q &= \int_{-\infty}^{\infty} u(t)s(t)\,dt \\ &= \frac{1}{2W} \sum_{j=-\infty}^{\infty} u(j/2W)s(j/2W) \\ &= \tilde{\mathbf{u}}\mathbf{s}, \end{aligned} \tag{7-9}$$

where **s** is a vector with components $s(j/2W)/\sqrt{2W}$.

Thus the total energy of a waveform is the square of the length of the vector representing it, and the cross-correlation of two waveforms is the scalar product of the vectors representing them.

**The Statistics of White Gaussian Noise.** A representation of the form (7-6) is particularly useful for white gaussian noise, i.e., gaussian noise whose power spectral density $\Psi(f)$ has a constant value, say $\frac{1}{2}N_0$, from $-W$ to $W$ and is zero elsewhere. Since $\Psi(f) = \frac{1}{2}N_0 \operatorname{rect}(f/2W)$, the correlation function of the noise is

$$\psi(\tau) = N_0 W \operatorname{sinc} 2W\tau, \tag{7-10}$$

which vanishes when $\tau$ is any multiple of $1/2W$ except zero. There it has the value $N_0W$, which is the total noise power or mean-squared noise voltage. Hence samples of this noise spaced in time by multiples of $1/2W$ are statistically independent zero-mean normal random variables with variance $N_0W$, and the components $u(j/2W)/\sqrt{2W}$ of the vector **u** representing the noise have a covariance matrix **M**, (1-23), consisting entirely of zeros except along the main diagonal, where its elements are all $\frac{1}{2}N_0$. We can write $\mathbf{M} = \frac{1}{2}N_0\mathbf{I}$, where **I** is the unit matrix, with ones on its main diagonal and zeros elsewhere.

To be able to use these statistics to determine the optimum receiver for discriminating between two specified band-limited signals, $s_1(t)$ and $s_0(t)$, in the presence of additive white gaussian noise, we must restrict the number $m$ of components of our vectors to a finite (but large) value, leaving for later the passage to the limit $m \to \infty$. Thus we suppose that the $m$ samples cover the entire interval during which the two signals differ significantly from each other, the omitted samples then being useless for distinguishing between the two signals. This interval should be such that $\int [s_1(t) - s_0(t)]^2\, dt$, when integrated over it, is very nearly equal to

$$E_{1-0} = \int_{-\infty}^{\infty} [s_1(t) - s_0(t)]^2\, dt, \tag{7-11}$$

and $\Sigma_j [s_1(j/2W) - s_0(j/2W)]^2/2W$, summed over the $m$ samples, is very close to $E_{1-0} = (\tilde{\mathbf{s}}_1 - \tilde{\mathbf{s}}_0)(\mathbf{s}_1 - \mathbf{s}_0)$, too. Later we shall pass to the limit $W \to \infty$ as well as $m \to \infty$, thereby removing the restriction to band-limited signals.

**Problem 7-4.** (*a*) Comparing (4-8), (4-13), and (4-14), notice that, if the narrow-band waveform (4-10) is confined to the bands from $\pm F - \frac{1}{2}B$ to $\pm F + \frac{1}{2}B$, then $x(t)$ and $y(t)$ contain no frequency higher than $\frac{1}{2}B$. Show, applying the foregoing sampling theorem to $x(t)$ and $y(t)$, that any band-limited narrow-band waveform (4-1) is completely determined by the samples $a(j/B)$ and $\phi(j/B)$ of its amplitude and phase for $j = \ldots, -1, 0, 1, \ldots$, being[1]

$$u(t) = \sum_{j=-\infty}^{\infty} a(j/B) \operatorname{sinc}(Bt - j) \cos [2\pi Ft + \phi(j/B)].$$

(*b*) Show that, if $u(t)$ is a narrow-band waveform confined to a band of width $B$ and $a(t)$ is its amplitude, then its total energy is

$$\int_{-\infty}^{\infty} u^2(t)\, dt = \frac{1}{2B} \sum_{j=-\infty}^{\infty} a^2(j/B).$$

**Problem 7-5.** Show that, if $u(t)$ is narrow-band white gaussian noise, with power spectral density constant over a band of width $B$, then the samples $a(j/B)$ and $\phi(j/B)(j = \ldots, -1, 0, 1, \ldots)$ of its amplitude and phase are all statistically independent, being Rayleigh- and uniformly distributed, respectively.

## ☐ THE OPTIMUM RECEIVER FOR TWO KNOWN SIGNALS

We now suppose that the waveform $u(t)$ available to the receiver, the "received signal," is either $s_1(t)$ or $s_0(t)$ plus white gaussian noise of power spectral density $\frac{1}{2}N_0$, that these two signals are completely known, and that they and the noise are confined to the band from $-W$ to $W$. Thus the vector $\mathbf{u}$, which by (7-6) is equivalent to $u(t)$, will have the statistics described on p. 119 except that its mean is not zero but is either $\mathbf{s}_1$ or $\mathbf{s}_0$,

[1] N. M. Blachman, A Comparison of the Informational Capacities of Amplitude- and Phase-Modulation Communication Systems, *Proc. IRE*, **41**:748–759 (1953), equation (A-5).

depending on which signal was added to the noise by the transmitter. The components of these two vectors are the signal samples $s_1(j/2W)$ and $s_0(j/2W)$ divided by $\sqrt{2W}$.

Restricting our attention to a large but finite number $m$ of components of $\mathbf{u}$ and noticing that the inverse of the covariance matrix $\mathbf{M} = \frac{1}{2}N_0\mathbf{I}$ is $\mathbf{M}^{-1} = (2/N_0)\mathbf{I}$, we obtain from (1-22) the likelihoods

$$p(\mathbf{u}|1) = (\pi N_0)^{-m/2} \exp\left[-\frac{(\tilde{\mathbf{u}} - \tilde{\mathbf{s}}_1)(\mathbf{u} - \mathbf{s}_1)}{N_0}\right],$$

$$p(\mathbf{u}|0) = (\pi N_0)^{-m/2} \exp\left[-\frac{(\tilde{\mathbf{u}} - \tilde{\mathbf{s}}_0)(\mathbf{u} - \mathbf{s}_0)}{N_0}\right].$$

Hence the logarithm of the likelihood ratio is, say,

$$\begin{aligned} L &= \ln \frac{p(\mathbf{u}|1)}{p(\mathbf{u}|0)} \\ &= \frac{(\tilde{\mathbf{u}} - \tilde{\mathbf{s}}_0)(\mathbf{u} - \mathbf{s}_0) - (\tilde{\mathbf{u}} - \tilde{\mathbf{s}}_1)(\mathbf{u} - \mathbf{s}_1)}{N_0} \\ &= \frac{2\tilde{\mathbf{u}}(\mathbf{s}_1 - \mathbf{s}_0) + \tilde{\mathbf{s}}_0\mathbf{s}_0 - \tilde{\mathbf{s}}_1\mathbf{s}_1}{N_0} \\ &= \frac{2q + E_0 - E_1}{N_0}, \end{aligned} \tag{7-12}$$

where $E_0$ and $E_1$ are the energies of the two signals and $q$ is the cross-correlation

$$q = \int_{-\infty}^{\infty} u(t)[s_1(t) - s_0(t)]\, dt. \tag{7-13}$$

In replacing the sums over $m$ components in (7-12) by the sums over *all* components, we passed to the limit $m \to \infty$, thereby obtaining the exact value of $L$ in place of our approximation.

Notice that the bandwidth $W$ does not appear in our result. It therefore is applicable to signals of any bandwidth whatever, finite or infinite, provided only that the spectrum of the noise is flat over all frequencies contained in either signal.

From (7-12) we see that $q$ is a monotonic function of the likelihood ratio and hence summarizes all of the information contained in $u(t)$ relevant to the decision between mark and space. An optimum receiver must therefore base its decisions on the value of (7-13) (or some monotonic function of it). (See Fig. 7-1.)

**The Matched Filter.** A receiver that evaluates $q$ by multiplying $u(t)$ by a locally generated or stored $s_1(t) - s_0(t)$ and integrating their product is called a *cross-correlating receiver*. A simpler method for evaluating (7-13) is based on the similarity of (7-13) to (3-3) for $t = T$, a time beyond which $s_1(t)$ and $s_0(t)$ no longer differ significantly. At time $T$ the

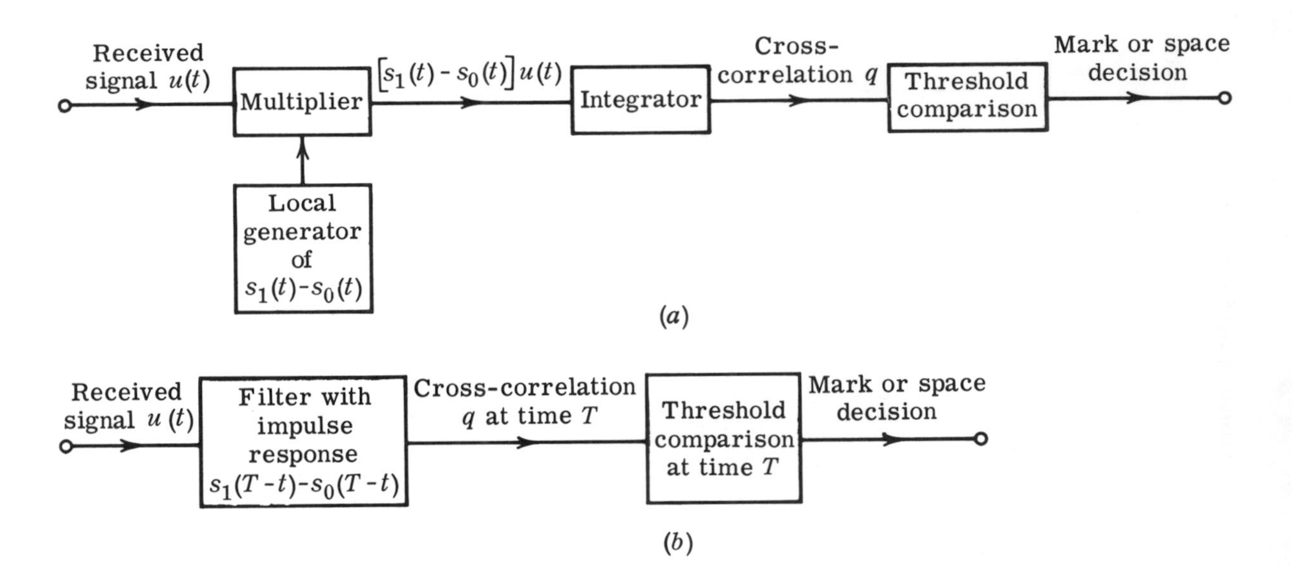


*Figure 7-1.* Two (equivalent) optimum receivers for distinguishing between two completely known signals in gaussian noise: (*a*) Cross-correlating recei·.er and (*b*) Matched-filter receiver.

output of a linear filter with impulse response $g(t)$ and input $u(t)$ is $\int_{-\infty}^{\infty} u(t)g(T - t)\,dt$, which will be $q$ if $g(T - t) = s_1(t) - s_0(t)$, that is, if $g(t) = s_1(T - t) - s_0(T - t)$.

Such a device is called a *matched filter;* its impulse response is the difference between the mark and space signals, reversed in time and delayed sufficiently (viz., by $T$) that $g(t)$ is zero for $t < 0$. Thus, for signals of finite duration, the matched filter is always realizable. In the case of on-off keying of a sinusoid, for example, where $s_0(t) = 0$ and $s_1(t)$ is a sinusoid of frequency $F$ and finite duration, a high-$Q$ $LC$ circuit resonant to this frequency can be used as a matched filter, provided that it is quenched at the time the mark signal begins and hence does not respond to earlier inputs. This quenching in effect terminates its sinusoidal impulse response, thereby making it $s_1(T - t)$ for all values of $t$ up to the time $T$ when the mark signal ends.

Besides being useful for optimum reception, the matched filter is distinguished by the fact that, among all linear filters, it has the highest output signal-to-noise ratio at the time $T$ when fed the difference signal $s(t) = s_1(t) - s_0(t)$ along with white gaussian noise.[1] The output-signal voltage at time $T$ of a filter with impulse response $g(t)$ is

$$E\left\{\int_{-\infty}^{\infty} u(T - t)g(t)\,dt\right\} = \int_{-\infty}^{\infty} s(T - t)g(t)\,dt,$$

and its average output-noise power is (3-12), which, by Parseval's theorem (3-11), can also be expressed as $\frac{1}{2}N_0 \int_{-\infty}^{\infty} g^2(t)\,dt$. The output signal-to-noise ratio at time $T$ is therefore

$$\frac{\left[\int_{-\infty}^{\infty} s(T - t)g(t)\,dt\right]^2}{\frac{1}{2}N_0 \int_{-\infty}^{\infty} g^2(t)\,dt}. \tag{7-14}$$

For each value of $t$ we set the derivative of (7-14) with respect to $g(t)$ equal to zero, getting

$$\frac{2s(T - t)}{\int_{-\infty}^{\infty} s(T - t)g(t)\,dt} = \frac{2g(t)}{\int_{-\infty}^{\infty} g^2(t)\,dt},$$

which tells us that $g(t)$ should be proportional to $s(T - t)$ in order to

[1] J. H. Van Vleck and D. Middleton, A Theoretical Comparison of the Visual, Aural, and Meter Reception of Pulsed Signals in the Presence of Noise, *J. Appl. Phys.*, **17**:940 (1946).

maximize[1] the output signal-to-noise ratio (7-14). The constant of proportionality does not affect (7-14); with $g(t) = s(T - t)$ it attains the value $2E_{1-0}/N_0$. Thus, just as the Wiener filter determined by (3-63) maximizes the signal-to-noise ratio when the input signal and noise are jointly ergodic with known correlation functions, the matched filter maximizes the signal-to-noise ratio for a completely known signal $s(t)$ plus noise with a known spectrum.

**Problem 7-6.** Show that, if a signal with Fourier transform $S(f)$ plus noise of power spectral density $\frac{1}{2}N(f)$ is fed to a linear filter having (complex) frequency response $G(f)$, its output signal-to-noise ratio at time $T$ will be maximum, namely, $2\int_{-\infty}^{\infty} |S(f)|^2\,df/N(f)$, when $G(f) = S^*(f)e^{-2\pi i f T}/N(f)$.

**Problem 7-7.** If the power spectral density $\frac{1}{2}N(f)$ of the received noise is not constant and is nowhere zero, a suitable filter at the receiver's input will whiten the noise, thus reducing the optimum-reception problem to the one we have already solved, since such a filter performs a reversible transformation that consequently preserves all of the information contained in the received signal. Show that the resulting optimum receiver is equivalent to the matched filter of Prob. 7-6 with $s(t) = s_1(t) - s_0(t)$.

**Geometric Interpretation.** The reason for the optimality of matched-filter reception can be understood by reference to the probability density function

$$p(\mathbf{u}|\boldsymbol{\mu}) = (\pi N_0)^{-m/2} \exp\left[-\frac{(\tilde{\mathbf{u}} - \tilde{\boldsymbol{\mu}})(\mathbf{u} - \boldsymbol{\mu})}{N_0}\right] \tag{7-15}$$

of the $m$-component vector $\mathbf{u}$ representing the received signal, whose mean $\boldsymbol{\mu}$ is either $\mathbf{s}_1$ or $\mathbf{s}_0$ accordingly as mark or space is transmitted. With each of the vectors $\mathbf{u}$ and $\boldsymbol{\mu}$ we associate a point of an $m$-dimensional euclidean space whose coordinates are the respective components of the vector. Thus $(\tilde{\mathbf{u}} - \tilde{\boldsymbol{\mu}})(\mathbf{u} - \boldsymbol{\mu})$, which is the sum of the squares of the respective component differences, is simply the square of the distance between the two points or, equivalently, the square of the length of the vector $\mathbf{u} - \boldsymbol{\mu}$.

Hence (7-15) depends only on the distance from the point $\boldsymbol{\mu}$ and is therefore spherically symmetric about $\boldsymbol{\mu}$, just as the bivariate normal

[1] With $h(t) = s(T - t)$, Schwarz's inequality

$$\left|\int_{-\infty}^{\infty} g(t)h(t)\,dt\right|^2 \le \int_{-\infty}^{\infty} |g(t)|^2\,dt \int_{-\infty}^{\infty} |h(t)|^2\,dt$$

confirms the fact that such a $g(t)$ maximizes (7-14). See the footnote to Prob. 3-28, and Wainstein and Zubakov, *op. cit.*, chap. 3, secs. 16 and 17.

distribution with equal variances and zero correlation coefficient, (1-17), is circularly symmetric about its mean. The problem of determining whether $s_1(t)$ or $s_0(t)$ was transmitted, therefore, is the problem of deciding whether $\mathbf{u}$ came from a population spherically symmetrically distributed about $\mathbf{s}_1$ or from another population identically distributed about $\mathbf{s}_0$.

Because of this symmetry, rotating the coordinate frame so that the $u_1$ axis is parallel to the vector $\mathbf{s}_1 - \mathbf{s}_0$ does not change the form of the distributions. Thus the new components of $\mathbf{u}$, like the old ones, have a spherically symmetric multivariate normal distribution and are, hence, likewise statistically independent of one another. Moreover, as a result of the rotation, all of the new components except $u_1$ have the same mean, whether $s_1(t)$ or $s_0(t)$ was transmitted. Having the same distribution for mark as for space, these components always yield a unit likelihood ratio and zero weight of evidence; they convey no useful information. Only the component of $\mathbf{u}$ in the direction of $\mathbf{s}_1 - \mathbf{s}_0$ is useful for discriminating between mark and space.[1]

This component is found by dividing the (scalar) product of $\mathbf{u}$ and $\mathbf{s}_1 - \mathbf{s}_0$, namely, $\tilde{\mathbf{u}}(\mathbf{s}_1 - \mathbf{s}_0)$, by the length of $\mathbf{s}_1 - \mathbf{s}_0$, namely, $\sqrt{(\tilde{\mathbf{s}}_1 - \tilde{\mathbf{s}}_0)(\mathbf{s}_1 - \mathbf{s}_0)}$. Referring to (7-11) and (7-13), we see that the information conveyed by $u(t)$ about which signal was sent is extracted by computing $u_1 = q/\sqrt{E_{1-0}}$ or, equally well, by determining $q$, since $E_{1-0}$ is a known constant. Hence the matched filter, which evaluates (7-13), evaluates the single useful component of the received signal $u(t)$.

**Problem 7-8.** The components $x$, $y$, and $z$ of the velocity of a gas molecule in three mutually perpendicular directions are independently normally distributed with mean zero and variance $T$ proportional to the temperature. Find and plot the distribution of the speed $v = \sqrt{x^2 + y^2 + z^2}$, which is known as the *Maxwell distribution*. For a two-dimensional gas, $v$ has a Rayleigh distribution. What would be its distribution for an $m$-dimensional gas? Show that in the latter case the expectation of $v^2$ is $mT$, while its standard deviation is $\sqrt{2m}\,T$, which becomes small compared to the expectation for large $m$. If follows that, while the direction of the velocity is uniformly distributed, its magnitude is nearly always relatively close to $\sqrt{mT}$ for large $m$.

**Problem 7-9.** From the fact that the total area under the probability density function of $v$ in Prob. 7-8 is unity, show that the surface area of a unit sphere in a space of $m$ dimensions is $A_m = 2\pi^{m/2}/\Gamma(\frac{1}{2}m)$ and that its volume is therefore $V_m = \pi^{m/2}/(\frac{1}{2}m)!$.

[1] See V. A. Kotel'nikov, "The Theory of Optimum Noise Immunity," sec. 3-3, p. 24, translated by R. A. Silverman, McGraw-Hill Book Company, New York, 1959.

**Error Probabilities.** Like the component of $\mathbf{u}$ in *any* direction, $u_1 = q/\sqrt{E_{1-0}}$ is normally distributed with variance $\frac{1}{2}N_0$, regardless of which signal is transmitted. However, its mean when mark is sent differs from its mean when space is sent by the length $\sqrt{E_{1-0}}$ of the vector $\mathbf{s}_1 - \mathbf{s}_0$, since $\mathbf{u}$ has mean $\mathbf{s}_1$ or $\mathbf{s}_0$, respectively, in the two cases. These two distributions of $u_1$ are shown in Fig. 7-2.

Since the logarithm of the likelihood ratio is a monotonic (in fact, a linear) function of $u_1$, any likelihood-ratio decision threshold divides the range of $u_1$ into two parts (Fig. 7-2). Below the threshold, space is the preferable decision and, above it, mark. The appropriate value for the threshold depends upon the prior probabilities of mark and space and upon the weight given to the various possible correct and incorrect

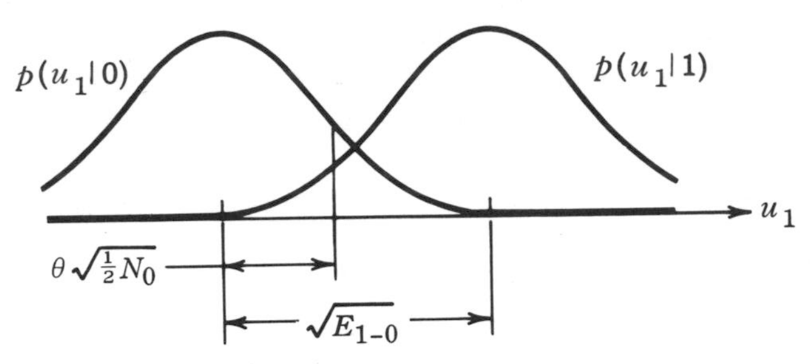


*Figure 7-2.* The distribution of $u_1$ when space is sent and when mark is sent.

decisions (Probs. 7-1 and 7-2), and the threshold, in turn, determines the error probabilities.

An "error of the first type" or "false alarm"—mistaking a space for a mark—will occur whenever a space is sent and $u_1$ exceeds the threshold, which we suppose is $\theta$ standard deviations above the mean of $p(u_1|0)$. Thus the false-alarm probability is $P_F = \text{erf}\,(-\theta)$ [see (1-5)]. An "error of the second type" or "miss"—mistaking a mark for a space—will occur when $u_1$ obeys the distribution $p(u_1|1)$ and is less than the threshold. Since the means of the two distributions are separated by $\sqrt{2E_{1-0}/N_0}$ standard deviations, its probability is

$$P_M = \text{erf}\,(\theta - \sqrt{2E_{1-0}/N_0}).$$

In the symmetric case, with equal prior probabilities for mark and space, equal credits for all correct decisions, and equal costs for all errors, the threshold should be located midway between the means of the two distributions, with $\theta = \sqrt{E_{1-0}/2N_0}$. The probabilities of errors of both types are then equal, being

$$P_F = P_M = \text{erf}\,(-\sqrt{E_{1-0}/2N_0}). \tag{7-16}$$

For on-off keying of a signal of energy $E_1$, the optimum receiver has an error probability in the symmetric case given by (7-16) with $E_{1-0} = E_1$. Note, however, that $E_{1-0}$ is *twice* the *average* signal energy. If the mark and space signals have equal energy $E_1 = E_0$ and are orthogonal, i.e., their cross-correlation (the integral of their product) is zero or, equivalently, the vectors $\mathbf{s}_1$ and $\mathbf{s}_0$ are perpendicular, by the pythagorean theorem we have $E_{1-0} = E_1 + E_0 = 2E_1$, which again is twice the (average) signal energy. This is the case, for example, with frequency-shift keying when the two frequencies used are separated by a multiple of the reciprocal of the signal duration.

On the other hand, if the two signals have equal energy $E_1 = E_0$ and are represented by antiparallel vectors, i.e., if $s_0(t) = -s_1(t)$, then $E_{1-0}$ is *four* times the (average) signal energy and yields the lowest possible error probability (7-16) for a given average signal energy. Notice that this minimum error probability is attained with $s_0(t) = -s_1(t)$, regardless of the form of $s_1(t)$, which can thus be selected on the basis of simplicity of implementation.

**Problem 7-10.** Prove that, for a given average signal energy $\frac{1}{2}E_1 + \frac{1}{2}E_0$, (7-16) is least when $\mathbf{s}_1$ and $\mathbf{s}_0$ are equal in length and opposite in direction.

**Problem 7-11.** Show that in the symmetric case the error probability of the optimum receiver of Prob. 7-7 is $\text{erf}\left[-\sqrt{\int_{-\infty}^{\infty} |S(f)|^2\, df/2N(f)}\right]$, where $S(f)$ is the Fourier transform of $s_1(t) - s_0(t)$ and $\frac{1}{2}N(f)$ is the power spectral density of the additive gaussian noise. Show that this becomes (7-16) for $N(f) = N_0$.

**Problem 7-12.** Show that, in the symmetric case, with matched-filter reception of equal-energy narrow-band signals in the presence of white gaussian noise of power spectral density $\frac{1}{2}N_0$, if the decision is based on the filter's output at a time differing by a very small amount $\tau$ from the time $T$ for which it was designed, the error probability becomes $\text{erf}\,(-\sqrt{E_{1-0}/2N_0}\cos 2\pi F\tau)$, where $F$ is any frequency in the band occupied by the signals.

## INCOHERENT RECEPTION

We have so far assumed that every detail of the two possible signals is precisely known at the receiver. The optimum detection of narrow-band signals under these circumstances is described as *coherent*. Narrow-band signals may, instead, be known precisely except for their phase; i.e., the mark and space signals may be

$$s_1(t) = a_1(t) \cos [2\pi Ft + \phi_1(t) + \theta_1]$$

and

$$s_0(t) = a_0(t) \cos [2\pi Ft + \phi_0(t) + \theta_0], \tag{7-17}$$

with $a_1(t)$, $\phi_1(t)$, $a_0(t)$, and $\phi_0(t)$ known but with $\theta_1$ and $\theta_0$ unknown constants having uniform prior distributions between 0 and $2\pi$ as the result of a small uncertainty concerning the exact value of the rapidly changing $2\pi Ft$ term (see Prob. 7-12)—possibly resulting from slow fluctuations in the length of the propagation path. In this case, antiparallel signals $s_0(t) = -s_1(t)$ cannot be distinguished.

Defining

$$
\begin{aligned}
s_{1c}(t) &= a_1(t) \cos [2\pi Ft + \phi_1(t)], & s_{1s}(t) &= a_1(t) \sin [2\pi Ft + \phi_1(t)], \\
s_{0c}(t) &= a_0(t) \cos [2\pi Ft + \phi_0(t)], & s_{0s}(t) &= a_0(t) \sin [2\pi Ft + \phi_0(t)],
\end{aligned} \tag{7-18}
$$

we have

$$s_1(t) = s_{1c}(t) \cos \theta_1 - s_{1s}(t) \sin \theta_1, \tag{7-19}$$

$$s_0(t) = s_{0c}(t) \cos \theta_0 - s_{0s}(t) \sin \theta_0, \tag{7-20}$$

or, in terms of the vectors representing these six waveforms,

$$
\begin{aligned}
\mathbf{s}_1 &= \mathbf{s}_{1c} \cos \theta_1 - \mathbf{s}_{1s} \sin \theta_1, \\
\mathbf{s}_0 &= \mathbf{s}_{0c} \cos \theta_0 - \mathbf{s}_{0s} \sin \theta_0.
\end{aligned}
$$

In general, the four vectors $\mathbf{s}_{1c}$, $\mathbf{s}_{1s}$, $\mathbf{s}_{0c}$, $\mathbf{s}_{0s}$ define a four-dimensional subspace of the signal space, and we suppose that the coordinate frame of the latter is rotated so that four of its axes define the same subspace. These four coordinates of the received-signal vector $\mathbf{u}$, then, have different statistics accordingly as mark or space is transmitted, but all of its other coordinates will have identical, independent, normal distributions in either case and hence do not affect the likelihood ratio.

To find the likelihood ratio, we therefore need only the probability density function of these four coordinates when mark is transmitted and when space is transmitted. If these two signals are orthogonal to each other for every $\theta_1$ and $\theta_0$, that is, if the four vectors $\mathbf{s}_{1c}$, $\mathbf{s}_{1s}$, $\mathbf{s}_{0c}$, $\mathbf{s}_{0s}$ are mutually perpendicular, as in the case of FSK with the two frequencies separated by a multiple of the reciprocal of the signal duration, the four coordinate axes can be parallel to these four vectors, but in general this is not possible. To evaluate the two likelihoods, it will be convenient to use different orientations of the four coordinate axes, which we shall call $u_1$, $u_2$, $u_3$, $u_4$ in one case and $u_{1'}$, $u_{2'}$, $u_{3'}$, $u_{4'}$ in the other.

To find the probability density per unit volume in this four-dimensional space, $p(u_1,u_2,u_3,u_4|1)$ when mark is transmitted, we choose the $u_1$ and $u_2$ axes parallel to $\mathbf{s}_{1c}$ and $\mathbf{s}_{1s}$. Then $u_3$ and $u_4$ are statistically independent of $u_1$ and $u_2$, and each contributes to the quadrivariate density function a factor of the form (1-3) with mean $\mu = 0$ and variance $\sigma^2 = \frac{1}{2}N_0$. To find $p(u_1,u_2|1)$, we notice that the transmitted signal (7-19) is represented by a vector $\mathbf{s}_1$ in the $u_1u_2$ plane of length $\sqrt{E_1}$, where

$$E_1 = \int_{-\infty}^{\infty} s_1^2(t)\,dt = \tfrac{1}{2}\int_{-\infty}^{\infty} a_1^2(t)\,dt = \int_{-\infty}^{\infty} s_{1c}^2(t)\,dt = \int_{-\infty}^{\infty} s_{1s}^2(t)\,dt \tag{7-21}$$

is the energy of the mark signal; because of the uniform distribution of $\theta_1$, its direction is uniformly distributed in this plane. Since the noise components in this plane have a circular normal distribution, the sum of the signal plus these noise components is also uniformly distributed in direction. Hence $p(u_1,u_2|1)$ depends only on the distance from the origin,

$$r_1 = \sqrt{u_1^2 + u_2^2}. \tag{7-22}$$

The probability density function of $r_1$ when a mark is sent does not depend on the phase $\theta_1$ of the signal, which we may thus suppose to be zero. Then $r_1$ has the same distribution as $a$ in Fig. 4-1 when the length $A$ of the signal phasor is $\sqrt{E_1}$ and the variance $\sigma^2$ of each noise component is $\frac{1}{2}N_0$. Substituting these values into (4-29), we get

$$p(r_1|1) = \frac{2r_1}{N_0} I_0\left(\frac{2\sqrt{E_1}\,r_1}{N_0}\right)\exp\left(-\frac{E_1 + r_1^2}{N_0}\right), \tag{7-23}$$

which, multiplied by $dr_1$, is the probability that the point $(u_1,u_2)$ falls within a ring of radius $r_1$ and thickness $dr_1$. Thus, dividing by the area of this ring, $2\pi r_1\,dr_1$, we get $p(u_1,u_2|1)$, and, multiplying by the univariate normal density functions for $u_3$ and $u_4$, we get

$$p(u_1,u_2,u_3,u_4|1) = \frac{1}{\pi^2 N_0^2} I_0\left(\frac{2\sqrt{E_1}\,r_1}{N_0}\right)\exp\left(-\frac{E_1 + r^2}{N_0}\right), \tag{7-24}$$

where

$$r = \sqrt{u_1^2 + u_2^2 + u_3^2 + u_4^2}$$

is the distance from the origin in our four-dimensional space.

To find $p(u_1,u_2,u_3,u_4|0)$, we use a rotated coordinate frame whose $u_{3'}$ and $u_{4'}$ axes are parallel to $\mathbf{s}_{0c}$ and $\mathbf{s}_{0s}$, respectively ($\mathbf{s}_{0c}$ and $\mathbf{s}_{0s}$, like $\mathbf{s}_{1c}$ and $\mathbf{s}_{1s}$, always being perpendicular to each other). By symmetry we then have

$$p(u_1,u_2,u_3,u_4|0) = \frac{1}{\pi^2 N_0^2} I_0\left(\frac{2\sqrt{E_0}\,r_0}{N_0}\right)\exp\left(-\frac{E_0 + r^2}{N_0}\right)$$

for the probability density per unit volume of the point called $(u_1,u_2,u_3,u_4)$ or $(u_{1'},u_{2'},u_{3'},u_{4'})$ when a space is sent, where

$$r_0 = \sqrt{u_{3'}^2 + u_{4'}^2} \tag{7-25}$$

and

$$E_0 = \int_{-\infty}^{\infty} s_0^2(t)\,dt = \tfrac{1}{2}\int_{-\infty}^{\infty} a_0^2(t)\,dt = \int_{-\infty}^{\infty} s_{0c}^2(t)\,dt = \int_{-\infty}^{\infty} s_{0s}^2(t)\,dt. \tag{7-26}$$

Hence the logarithm of the likelihood ratio is

$$L = \ln I_0\left(\frac{2\sqrt{E_1}\,r_1}{N_0}\right) - \ln I_0\left(\frac{2\sqrt{E_0}\,r_0}{N_0}\right) - \frac{E_1 - E_0}{N_0}, \qquad (7\text{-}27)$$

which depends upon the received signal $u(t)$ only through $r_1$ and $r_0$. Notice that $r_1{}^2$ and $r_0{}^2$ can be regarded as the total mark energy received and the total space energy received, respectively, including a contribution from the noise whose expected value in each case is $N_0$, since $r_1$ is the length of the projection of $\mathbf{u}$ onto the mark-signal plane and $r_0$ is the length of its projection onto the space-signal plane.

**Optimum Receiver.** If the space signal is not a constant multiple of the mark signal, i.e., if the space-signal plane does not coincide with the mark-signal plane, the value of $r_1$ does not determine the value of $r_0$. Hence an optimum receiver must calculate $r_1$ and $r_0$ (or monotonic functions of them) in order that its decisions may be based on the value of (7-27) (or some monotonic function of it).

In principle, $r_1$ and $r_0$ can be obtained from the received signal $u(t)$ by means of four cross-correlators (Fig. 7-3*a*), which multiply $u(t)$ by each of the four waveforms (7-18), integrating, and taking the square roots (7-22) and (7-25). The matched-filter technique, however, is much more easily implemented, for

$$s_{1c}(t) = s_{1s}\left(t + \frac{1}{4F}\right)$$

and hence a filter that is matched to $s_{1c}(t)$ with a delay $T$, that is, with impulse response $s_{1c}(T - t)$, is matched to $s_{1s}(t)$ with a delay $T + 1/4F$. Since $s_{1c}(t)$ oscillates sinusoidally with frequency $F$, the filter's output oscillates similarly. Because $\cos^2\phi + \sin^2\phi = 1$, the amplitude of this oscillation equals the square root of the sum of the squares of the instantaneous output values a quarter of a cycle apart, and around time $T$ is therefore proportional to $r_1$, being in fact $\sqrt{E_1}\,r_1$.

Hence an optimum receiver can be realized (Fig. 7-3*b*) by feeding the received waveform $u(t)$ to two filters—one matched to the mark signal and the other to the space—whose outputs are fed to two rectifiers and wide low-pass filters that integrate over a few cycles of the frequency $F$. If these rectifiers are such as to yield the logarithms of Bessel functions appearing in (7-27) at the low-pass-filter output (see Prob. 5-2), the decision between mark and space can be based simply on the difference of their outputs after low-pass filtering.

This result could have been derived from our result for the coherent detection of the same two signals, since the only difference between the two cases is, in effect, a small uncertainty as to the epoch of the signal in incoherent reception. For each of the two signals and each possible epoch, the corresponding cross-correlation or matched-filter output at time $T$ indicates the likelihood of the presence of a signal with that epoch. Thus the matched-filter outputs over a short interval around $T$ give all of the available information about the presence of a signal with its epoch in the interval of our uncertainty. For a narrow-band signal the cross-correlation or matched-filter output oscillates sinusoidally over any short time interval and can be described completely by its amplitude and phase. The phase evidently carries the information about the epoch of a signal, and the amplitude conveys the information about its presence, as we have just seen.[1] Because of their complexity or cost, it will not always be worthwhile to construct filters matched to given signals, but optimum receivers merit our interest because they provide a standard of comparison for practical receivers, showing how much is sacrificed in the way of performance and indicating a level that cannot be surpassed, regardless of effort or expense.

In the case of frequency-shift keying, each matched filter can simply be a high-$Q$ $LC$ circuit tuned to one of the two signal frequencies, both resonant circuits being quenched at the time the signals begin so that they are not responding to earlier inputs. If they are quenched again at the time $T$, when the signals have ended, the receiver will be ready to receive another mark or space. Thus a very natural sort of receiver for FSK is actually optimum.

If one signal is zero or is a constant multiple of the other, $r_0 = r_1$, and only a single matched filter is needed (Fig. 7-3c). In this case the decision can be based on the value of *any* monotonic function of $r_1$, and no special rectifier characteristic or nonlinearity is needed.

**Problem 7-13.** Devise an optimum receiver for orthogonal signals (7-17) with constant amplitude over the signal duration, with phase $\phi_0(t) = 0$ throughout, and with phase $\phi_1(t)$ equal to zero for the first half of the signal duration and $\pi$ for the second half, there being a short transition interval in the middle to keep the signal narrow-band. Since information is conveyed here by the change of phase (or lack of it) rather than by the phase itself, the use of such signals is called *differential-phase-shift keying,* and their reception is *differentially coherent.* Show that the optimum receiver in effect compares the phase of the response of an $LC$ circuit resonant to frequency $F$ which is

[1] When the uncertainty as to the signal's epoch covers a time that is not short compared to the reciprocal of the signal's bandwidth, we must deal with likelihoods that are averaged over more than just a cycle of oscillation of the cross-correlation, and it is sometimes important not only to determine *whether* a signal is present but also *when* it began, as in the case of radar. This is the "parameter-estimation problem." See P. M. Woodward, "Probability and Information Theory with Applications to Radar," chap. 5, Pergamon Press, New York, 1953; Carl W. Helstrom, "Statistical Theory of Signal Detection," chaps. VIII and IX, Pergamon Press, New York, 1960.

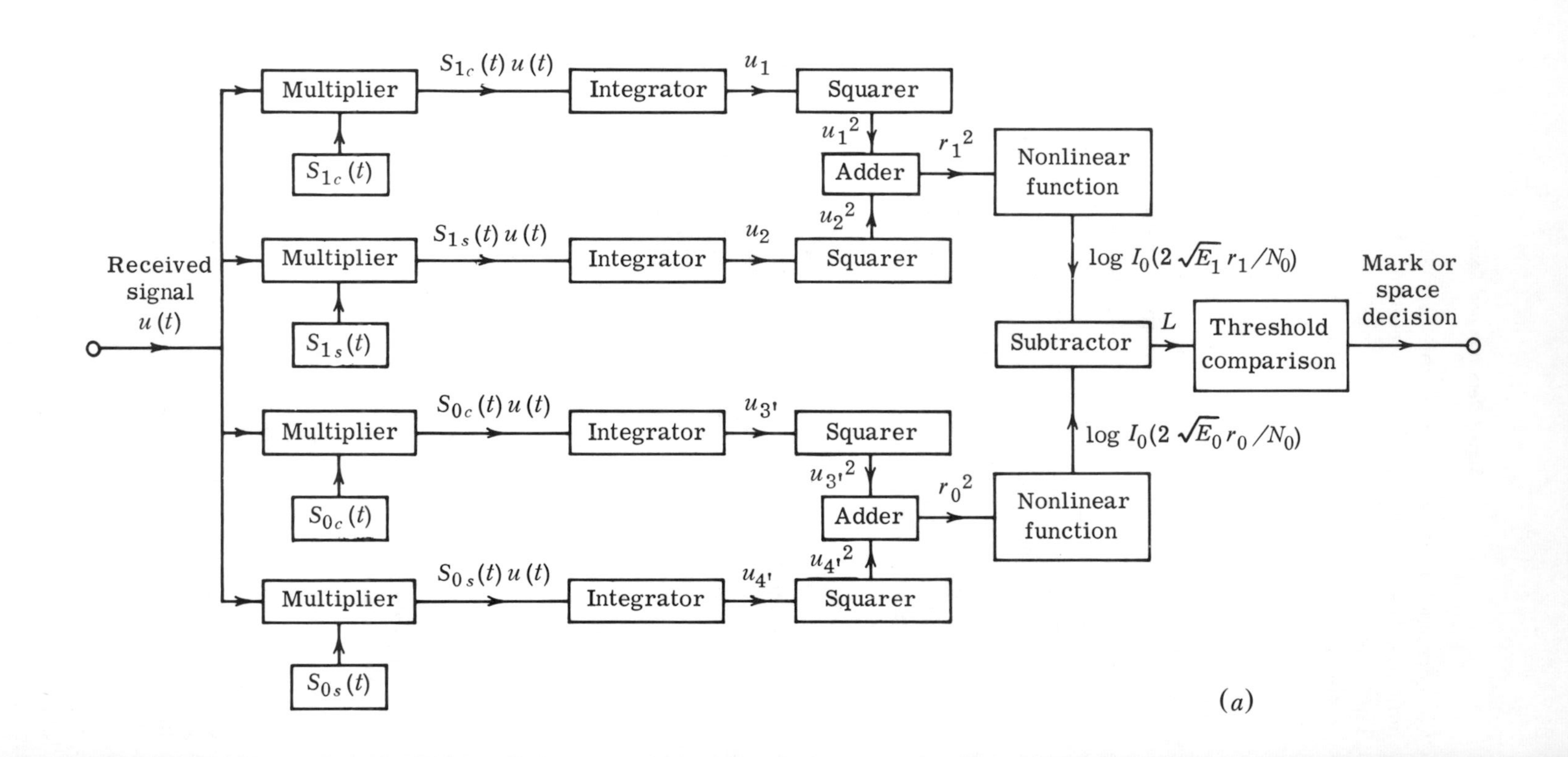
Received
signal
u(t)
Multiplier
S1c(t)
S1c(t) u(t)
Integrator
u1
Squarer
u1²
Adder
r1²
Nonlinear
function
Multiplier
S1s(t)
S1s(t) u(t)
Integrator
u2
Squarer
u2²
log I0(2 √E1 r1/N0)
Subtractor
L
Threshold
comparison
Mark or
space
decision
Multiplier
S0c(t)
S0c(t) u(t)
Integrator
u3'
Squarer
u3'²
Adder
r0²
Nonlinear
function
log I0(2 √E0 r0/N0)
Multiplier
S0s(t)
S0s(t) u(t)
Integrator
u4'
Squarer
u4'²

(a)

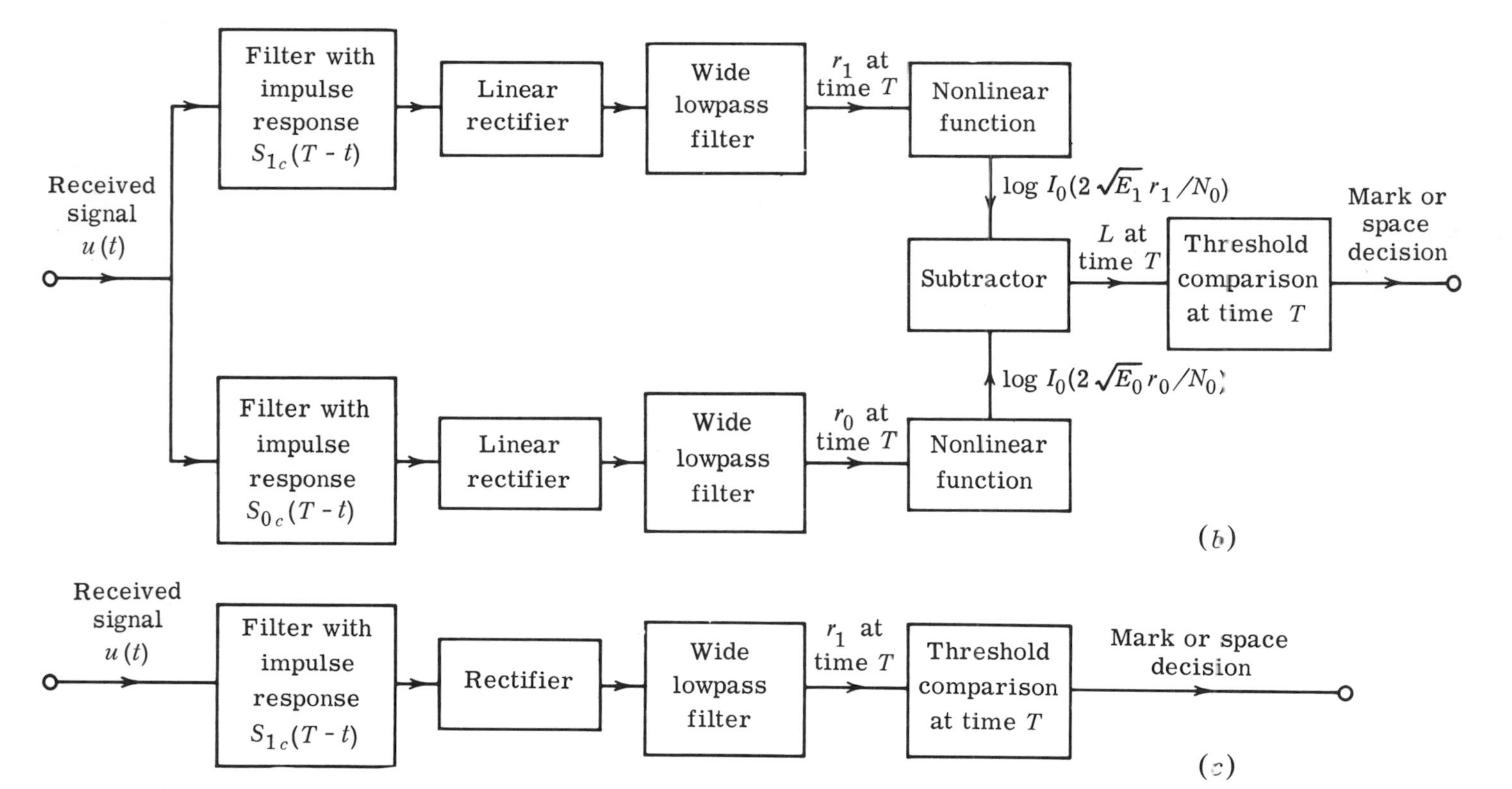


*Figure 7-3.* Two (equivalent) optimum incoherent receivers for distinguishing between two narrow-band signals of known form but unknown phase in gaussian noise, (*a*) cross-correlating receiver and (*b*) matched-filter receiver, and (*c*) matched-filter receiver for the case in which one signal is either identically zero or is a multiple of the other signal.

fed $u(t)$ during the first half of the signal with the phase of the output of a similar circuit fed $u(t)$ during the second half, deciding in favor of space when the phase difference is less than 90° if the decision criterion is the sign of the weight of evidence (7-27).

**Problem 7-14.** Show that, in the case of incoherent reception of a signal that can take more than two different forms, the use of a filter matched to each of them, followed by a rectifier and wide low-pass filter, will yield a set of voltages that suffice to determine the posterior probabilities of the various signals. Hence these filters and rectifiers can form part of an optimum receiver, which extracts all of the pertinent information from the received signal. Notice that it may sometimes be possible to use fewer matched filters. What modifications are necessary if the noise is not white?

**Problem 7-15.** Describe an optimum coherent receiver for more than two different signals in the presence of gaussian noise.

**Error Probability for Equal-Energy Orthogonal Signals.** We now suppose that the mark and space signals have equal prior probabilities and equal importance, and that therefore the optimum receiver bases its decisions on the sign of the weight of evidence (7-27) in favor of mark contained in the received signal. If both signals have equal energy, as in FSK or differentially coherent keying (Prob. 7-13), that is, if (7-21) equals (7-26), the sign of $L$ is positive or negative accordingly as $r_1 > r_0$ or $r_1 < r_0$. Here again no special rectifier characteristics or nonlinearities are necessary for optimum reception.

If, in addition, the mark and space signals are orthogonal, as in the case of differentially coherent keying or FSK with the frequency difference equal to a multiple of the reciprocal of the signal duration, the $u_1$ and $u_2$ axes are perpendicular to the $u_{3'}$ and $u_{4'}$ axes. Hence the noise components affecting $r_1$ are statistically independent of those that affect $r_0$, and consequently $r_1$ and $r_0$ are independent of one another, that is, $p(r_1,r_0|1) = p(r_1|1)p(r_0|1)$, for example. As a result, it is easy to calculate the error probability.

In this case the probability that a mark will be mistaken for a space is the probability that $r_0 > r_1$ when $r_1$ has the distribution (7-23), and $r_0$ has the Rayleigh distribution

$$p(r_0|1) = \frac{2r_0}{N_0} \exp\left(-\frac{r_0^2}{N_0}\right),$$

obtained from (7-23) by replacing $r_1$ by $r_0$ and $E_1$ by zero. For any given value of $r_1$, the probability that $r_0 > r_1$ is the integral of this Rayleigh density function from $r_1$ to $\infty$, namely, $\exp(-r_1^2/N_0)$, which we must average over the distribution (7-23) of $r_1$. Multiplying it by (7-23) and factoring out $\frac{1}{2}\exp(-E_1/2N_0)$, we see that the remaining factor has the same form as (7-23) except that $E_1$ and $r_1$ have been replaced by $\frac{1}{2}E_1$ and $\sqrt{2}\, r_1$, respectively. Thus the integral of this factor from zero to infinity, like that of (7-23), is unity, and the error probability is

$$P_M = P_F = \tfrac{1}{2} \exp\left(-\frac{E_1}{2N_0}\right), \tag{7-28}$$

being the same for errors of both types because of the symmetry.

In applying this result to differentially coherent reception (Prob. 7-13), it should be kept in mind that $E_1$ is *twice* the energy of the second half of the signal, which can serve as the first half of another similar signal.

**Problem 7-16.** Generalize (7-28) to the case of colored noise.

**Problem 7-17.** Show that, if decisions are based on the sign of $L$, the error probability with optimum incoherent reception of slowly Rayleigh fading equal-energy orthogonal binary signals of average energy $E$ in the presence of additive white gaussian noise of power spectral density $\frac{1}{2}N_0$ is $N_0/(E + 2N_0)$. Here the signal (but not the noise) is multiplied by a Rayleigh-distributed factor which is constant over the duration of the signal. Note that the optimum receiver is here independent of this factor and that the energy of the received signal obeys an exponential distribution (see Prob. 3-4). Such fading can result from severe multipath propagation.

**Problem 7-18.** (*a*) By expressing $p(r_1,r_0|1)$ as $p(r_1|1)p(r_0|r_1, 1)$, show that, when the mark and space planes are not orthogonal, but instead make an angle $\alpha$ whose cosine is

$$\cos \alpha = \frac{1}{2\sqrt{E_1E_0}} \left| \int_{-\infty}^{\infty} a_1(t)a_0(t)e^{i\phi_1(t) - i\phi_0(t)}\, dt \right|$$

(that is, every vector in either plane makes angle $\alpha$ with the other plane, the origin being the only point common to the two planes), we have[1]

$$p(r_1,r_0|1) = \frac{4r_1r_0}{N_0^2 \sin^2 \alpha} I_0\left(\frac{2\sqrt{E_1}\, r_1}{N_0}\right) I_0\left(\frac{2r_1r_0 \cos \alpha}{N_0 \sin^2 \alpha}\right) \exp\left(-\frac{E_1}{N_0} - \frac{r_1^2 + r_0^2}{N_0 \sin^2 \alpha}\right).$$

[1] C. W. Helstrom, The Resolution of Signals in White Gaussian Noise, *Proc. IRE*, **43**:1111–1118 (1955), has evaluated the probability that $r_0 > r_1$ for this distribution, namely,

$$P_M = Q\left(\sqrt{\frac{E_1}{N_0}} \sin \frac{\pi - 2\alpha}{4}, \sqrt{\frac{E_1}{N_0}} \cos \frac{\pi - 2\alpha}{4}\right) - \tfrac{1}{2} \exp\left(-\frac{E_1}{2N_0}\right) I_0\left(\sqrt{\frac{E_1}{2N_0}} \cos \alpha\right)$$

with the $Q$ function defined by (7-30), which is the probability of mistaking mark for space if the decision is based on whether $r_1$ or $r_0$ is larger. This is the optimum criterion when both signals have the same energy, prior probability, and importance.

(b) When mark and space have equal importance and equal prior probabilities, for a given average signal energy $\frac{1}{2}E_1 + \frac{1}{2}E_0$, do equal-energy orthogonal signals minimize the minimum average error probability $\frac{1}{2}P_F + \frac{1}{2}P_M$, minimized over all choices of the threshold, for optimum incoherent reception?

**Error Probability when $E_0 = 0$.** In general, the evaluation of the error probabilities for optimum incoherent reception requires the integration of $p(r_1,r_0|1)$ over the part of the $r_1r_0$ plane (Fig. 7-4) in which the likelihood ratio is less than the optimum threshold, i.e., the region above the corresponding likelihood-ratio contour, to get $P_M$, and the integration of $p(r_1,r_0|0)$ over the rest of the plane to get $P_F$, the probability of mistaking a space for a mark. One contour is a straight line of slope $\sqrt{E_1/E_0}$ through the origin, and the others are asymptotically parallel to it. Except in the case of the straight-line contour, as in the preceding section, these integrations are unpleasant, even when the mark and space signals are orthogonal, though they can be carried out numerically if necessary.

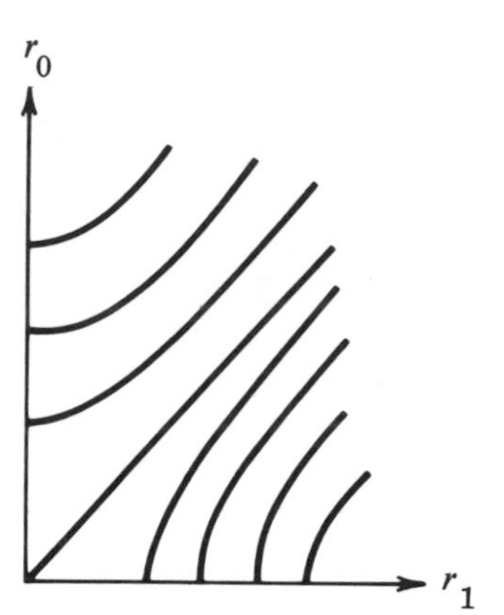


*Figure 7-4.* Likelihood-ratio contours.

However, some simplification results when $E_0 = 0$ and the "space" signal is actually just a space, for (7-27) then depends only on $r_1$, and optimum decisions can be based simply on whether $r_1$ exceeds some appropriate threshold, say $\sqrt{\frac{1}{2}N_0}\,\theta$. The probability that a mark will be mistaken for a space can then be written as

$$P_M = 1 - \int_{\sqrt{\frac{1}{2}N_0}\theta}^{\infty} \frac{2r_1}{N_0} I_0\left(\frac{2\sqrt{E_1}\,r_1}{N_0}\right) \exp\left(-\frac{E_1 + r_1^2}{N_0}\right) dr_1$$
$$= 1 - Q(\sqrt{2E_1/N_0},\, \theta), \tag{7-29}$$

where[1]

$$Q(R,\theta) = \int_\theta^\infty rI_0(Rr) \exp\left(-\frac{R^2 + r^2}{2}\right) dr \tag{7-30}$$

is the complement of the cumulative distribution function corresponding to the probability density (4-29) with $\sigma = 1$. Accordingly, $Q(R,0) = 1$ for any $R$ and $Q(R,\infty) = 0$, but $Q(\infty,\theta) = 1$ for any finite $\theta$. For $\theta = R$ we have $Q(R,R) = \frac{1}{2} + \frac{1}{2}I_0(R^2)e^{-R^2}$.

The probability of false alarm, i.e., the probability that $r_1$ will ex-

[1] J. I. Marcum, A Statistical Theory of Target Detection by Pulsed Radar, *IRE Trans. Inform. Theory,* **IT-6**:59–267 (April 1960); and "A Table of the $Q$ Function," Research Memorandum RM-339, Rand Corporation, Santa Monica, Calif., January, 1950.

ceed the threshold in the absence of a signal, is obtained by setting $E_1 = 0$ in the integral of (7-29), namely,

$$\begin{aligned} P_F &= \int_{\sqrt{\frac{1}{2}N_0}\theta}^{\infty} \frac{2r_1}{N_0} \exp\left(-\frac{r_1^2}{N_0}\right) dr_1 \\ &= Q(0,\theta) \\ &= e^{-\frac{1}{2}\theta^2}. \end{aligned} \tag{7-31}$$

Thus it is easy to determine the threshold that gives any specified false-alarm probability and to find the corresponding miss probability (7-29). However, it may be more important to minimize the average error probability $\frac{1}{2}P_F + \frac{1}{2}P_M$ than to obtain a specified $P_F$.

We have supposed that the received signal $u(t)$ is known for all $t$. Optimum detection can also be studied in the case where $u(t)$ can be observed for only a finite or semi-infinite time.[1] In that case $u(t)$ is expanded as a Karhunen-Loève series, in terms of the eigenfunctions of an integral equation whose kernel is the correlation function of the noise. In general these eigenfunctions cannot be found analytically, and useful results can be obtained only when the noise spectrum is rational.[2] For this reason we shall not go further into the matter here.

**Problem 7-19.** Show that $Q(R,\theta) \cong \text{erf}\,(R - \theta)$ for $R \gg 1 + |R - \theta|$.

## ☐ DETECTION OF NOISE

We now turn from signals which are completely known except for their phase to signals which are completely unknown except for their spectra. We suppose that the received signal $u(t)$, which includes the inevitable background noise, is a sample of a zero-mean ergodic gaussian random process with power spectral density either $\Psi_0(f)$ or $\Psi_1(f)$. This situation arises, for example, when looking for a radio star with a radio telescope or when communicating by means of known binary signals through a medium whose phase shift varies wildly and unpredictably with frequency.

**Instantaneous Observation.** If $u(t)$ is observed long enough, its power spectrum can be determined arbitrarily well, and an arbitrarily reliable decision can be made as to whether it is $\Psi_0(f)$ or $\Psi_1(f)$. We therefore suppose that a decision must be made after observing $u(t)$ for a time

[1] Carl W. Helstrom, "Statistical Theory of Signal Detection," chaps. IV and V, Pergamon Press, New York, 1960; Ivan Selin, "Detection Theory," chap. 7, Princeton University Press, Princeton, N.J., 1965.

[2] D. Middleton, "An Introduction to Statistical Communication Theory," chap. 8 and app. 2, McGraw-Hill Book Company, New York, 1960; W. B. Davenport, Jr., and W. L. Root, "An Introduction to the Theory of Random Signals and Noise," app. 2, McGraw-Hill Book Company, New York, 1958.

$T$. If $T$ is extremely short, the problem is to decide whether $u = u(0)$, say, is drawn from a zero-mean normal population with variance

$$\sigma_1^2 = \int_{-\infty}^{\infty} \Psi_1(f)\, df \qquad \text{or} \qquad \sigma_0^2 = \int_{-\infty}^{\infty} \Psi_0(f)\, df.$$

The conditional probability density functions $p(u|1)$ and $p(u|0)$ are given by (1-3) with $\mu = 0$. Since the ratio of these two likelihoods,

$$\frac{p(u|1)}{p(u|0)} = \frac{\sigma_0}{\sigma_1} \exp\left(\frac{u^2}{2\sigma_0^2} - \frac{u^2}{2\sigma_1^2}\right),$$

is a monotonic function of $u^2$, the decision should be based on whether $u^2$ exceeds some appropriate threshold, say $\theta^2$, that is, on whether $|u| > \theta$. If $\sigma_1 > \sigma_0$, the probability of mistaking $\Psi_0(f)$ for $\Psi_1(f)$ is the probability that $|u| > \theta$ when $u$ has the probability density function $p(u|0)$, viz., by (1-5),

$$P_F = 2 \operatorname{erf}\left(-\frac{\theta}{\sigma_0}\right). \tag{7-32}$$

The probability of deciding in favor of $\Psi_1(f)$ when it is actually present, on the other hand, is the probability that $|u| > \theta$ when $u$ has the variance $\sigma_1^2$, namely,

$$P_D = 2 \operatorname{erf}\left(-\frac{\theta}{\sigma_1}\right). \tag{7-33}$$

The *detection probability* $P_D$ is the complement of the probability of an error of the second type, $1 - P_M$.

Either of these probabilities, $P_F$ or $P_D$, can be made to take any desired value by an appropriate choice of the threshold $\theta$, and the other will then be determined. The lower $P_F$ is made, however, the lower $P_D$ will be. A graph of $P_D$ as a function of $P_F$, with $\theta$ as a parameter, is known as a *receiver operating characteristic* (ROC). Solving (7-32) for $\theta$ and substituting into (7-33), we see that its equation is here

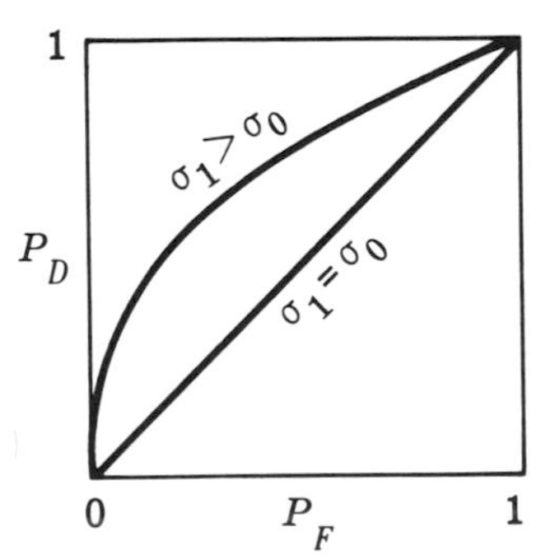


*Figure 7-5.* Receiver operating characteristics.

$$P_D = 2 \operatorname{erf}\left[\frac{\sigma_0}{\sigma_1} \operatorname{erf}^{-1} \left(\tfrac{1}{2}P_F\right)\right], \tag{7-34}$$

where $\operatorname{erf}^{-1}$ is the inverse of the error function (1-5). When $\sigma_0 = \sigma_1$, (7-34) becomes $P_D = P_F$, and the ROC is the straight line joining (0,0) and (1,1) (Fig. 7-5), for in this case $u$ conveys no information, and the decisions must be random guesses of $\Psi_1(f)$ with probability $P_F$. Otherwise, the ROC bows upward to give $P_D > P_F$ except at the end points (0,0) and (1,1).

**Short Observation of Narrow-Band Noise.** When $u(t)$ is a narrow-band random process and the observation interval $T$ is about one cycle of its center frequency, its instantaneous amplitude $a$ and phase $\phi$ can be determined. Being statistically independent of $a$ and uniformly distributed in any case, the phase does not affect the likelihood ratio and gives no information about the signal. Since the amplitude obeys the Rayleigh distribution

$$p(a|1) = \frac{a}{\sigma_1^2} \exp\left(-\frac{a^2}{2\sigma_1^2}\right) \quad \text{or} \quad p(a|0) = \frac{a}{\sigma_0^2} \exp\left(-\frac{a^2}{2\sigma_0^2}\right),$$

the likelihood ratio is

$$\frac{p(a|1)}{p(a|0)} = \frac{\sigma_0^2}{\sigma_1^2} \exp\left(\frac{a^2}{2\sigma_0^2} - \frac{a^2}{2\sigma_1^2}\right),$$

which is a monotonic function of $a$, inasmuch as $a \geq 0$.

The decision should therefore be made according to whether $a$ exceeds some appropriate threshold $\theta$. The resulting false-alarm and detection probabilities are given by the integral of the Rayleigh distribution of $a$ from $\theta$ to $\infty$,

$$P_F = \exp\left(-\frac{\theta^2}{2\sigma_0^2}\right) \quad \text{and} \quad P_D = \exp\left(-\frac{\theta^2}{2\sigma_1^2}\right). \tag{7-35}$$

Eliminating $\theta$, we find that the equation of the ROC is here

$$P_D = P_F^{\sigma_0^2/\sigma_1^2}. \tag{7-36}$$

This exceeds (7-34) except at the end points (0,0) and (1,1) if $\sigma_1 > \sigma_0$.

**Problem 7-20.** What is the equation for the ROC of an optimum coherent receiver for two known signals in the presence of additive white gaussian noise?

**Problem 7-21.** What can be done to improve a receiver whose ROC falls below the line $P_D = P_F$? How will this improvement affect its ROC?

**Problem 7-22.** Show that the slope of the ROC at any point equals the likelihood ratio at that point.

**Extrapolation of Band-Limited Waveforms.** If $u(t)$ is a band-limited waveform whose values we know throughout some interval, we can differentiate it any number of times to obtain coefficients for its Taylor series expansion. Since the rate of growth of the derivatives with their order is limited by the cutoff frequency, the radius of convergence of the Taylor series is infinite, and it enables us to find $u(t)$ for *any* $t$. Thus an observation interval of finite length is equivalent to an infinite one.

However, it is not practical to extrapolate $u(t)$ very far in this way, for it requires many terms of the Taylor series and hence both taxes the ability to measure $u(t)$ accurately in order to determine its higher derivatives, and strains the credibility of band-limiting, as each differentiation increases the weight given to higher frequencies as compared to those below the band limit.

**Long Observation.** The case in which the duration $T$ of the observation of $u(t)$ is of the order of the reciprocal of its bandwidth[1] presents difficulties which are related to the fact that $u(t)$ can be extrapolated over such a time both before and after the observation interval. Thus we pass on to the case of very large $T$, which is particularly important when the two noise spectra differ relatively little. Here $u(t)$, which need not be narrow-band, can be represented during the observation interval by the Fourier series

$$u(t) = \frac{A_0}{\sqrt{T}} + \sqrt{\frac{2}{T}} \sum_{j=1}^{\infty} \left( A_j \cos \frac{2j\pi t}{T} - B_j \sin \frac{2j\pi t}{T} \right), \tag{7-37}$$

which is not meant to suggest that $u(t)$ is periodic. The terms of (7-37) are orthogonal to each other over the interval $(0,T)$; the factors $1/\sqrt{T}$ and $\sqrt{2/T}$ give each term, apart from its coefficient, a unit energy. If frequencies less than $W$, which we shall later allow to become infinite, account for nearly all of the power in the spectrum of $u(t)$, we can truncate the Fourier series (7-37) after $j = WT$, that is, after $2WT + 1$ terms, and incur an arbitrarily small mean-squared error. [If $T$ is not large, however, the spectrum of (7-37) will be determined more by end effects than by the behavior of $u(t)$ itself.] Thus $u(t)$ not only determines, but also is determined by, the set of $2WT + 1$ Fourier coefficients $(A_0, A_1, \ldots, A_{WT}, B_1, \ldots, B_{WT})$, which therefore constitute a set of coordinates for the vector $\mathbf{u}$ representing $u(t)$. In terms of these coordinates and $B_0 = 0$,

[1] This case can be treated in a manner related to that used for finite-duration observations of known signals in gaussian noise. See Middleton, *op. cit.*, sec. 20.4-7; Helstrom, *op. cit.*, chap. XI; Selin, *op. cit.*, chap. 8; T. T. Kadota, Optimum Reception of Binary Gaussian Signals, *Bell System Tech. J.*, **43**:2767–2810 (1964). T. T. Kadota, Optimum Reception of Binary Sure and Gaussian Signals, *Bell System Tech. J.*, **44**:1621–1658 (1965), and Optimum Reception of $M$-ary Gaussian Signals in Gaussian Noise, *Bell System Tech. J.*, **44**:2187–2197 (1965), has elegantly combined the problem of known signals observed for a finite time with that of gaussian signals having known covariance functions (or spectra) by treating the known signals as the (varying) mean values of nonstationary gaussian random processes. The results are expressed in terms of the solutions of suitable integral equations.

the energy of $u(t)$ in the interval $(0,T)$ is

$$E_u = \int_0^T u^2(t)\,dt = \sum_{j=0}^{WT} (A_j^2 + B_j^2). \tag{7-38}$$

By reference to (7-8) we see that $E_u$ can also be approximated for large $T$ by the summation

$$E_u = \sum_{j=0}^{2WT} u_j^2 \tag{7-39}$$

over the $2WT + 1$ samples $u_j = u(j/2W)/\sqrt{2W}$ of $u(t)$ falling within the observation interval. Since the square of the length of the vector $\mathbf{u}$ has the pythagorean form (7-38) or (7-39) in terms of either the sampling-theorem coordinates $(u_0, u_1, \ldots, u_{2WT})$ or the Fourier coefficients $(A_0, \ldots, B_{WT})$, it follows[1] that the difference between these two sets of coordinates is simply the result of a rotation of the coordinate frame (and, possibly, a reflection). This rotation thus changes the coordinates from the time domain $(u_0, \ldots, u_{2WT})$ to the frequency domain $(A_0, \ldots, B_{WT})$.

If $u(t)$ were white gaussian noise of spectral density $\frac{1}{2}N_0$, as (7-10) indicates, its time-domain coordinates would be independent zero-mean normal random variables with variance $\frac{1}{2}N_0$. Since the resulting joint probability density is spherically symmetric about the origin, a rotation of the coordinate frame does not alter its form, and hence the Fourier coefficients would similarly be independent, zero-mean, normal random variables with variance $\frac{1}{2}N_0$. However, $u(t)$ is not white noise but has power spectral density $\Psi_0(f)$ or $\Psi_1(f)$, as if it had resulted from passing white gaussian noise through a filter with frequency response $|G(f)|$ proportional to $\sqrt{\Psi_0(f)}$ or $\sqrt{\Psi_1(f)}$ [see (3-12)]. The effect of this filtering on (7-37) is to change the variance of $A_j$ and $B_j$ to $\Psi_0(j/T)$ or $\Psi_1(j/T)$, leaving the Fourier coefficients still independent zero-mean normal random variables (see Prob. 7-23).

Since the probability density function of each Fourier coefficient is given by (1-3), with $\mu = 0$ and $\sigma^2 = \Psi_0(j/T)$ or $\Psi_1(j/T)$, the logarithm of the likelihood ratio $p(\mathbf{u}|1)/p(\mathbf{u}|0)$ is

$$L = \frac{1}{2} \sum_{j=0}^{WT} \left[ \frac{1}{\Psi_0(j/T)} - \frac{1}{\Psi_1(j/T)} \right] (A_j^2 + B_j^2) - \frac{1}{2} \sum_{j=-WT}^{WT} \ln \frac{\Psi_1(j/T)}{\Psi_0(j/T)}. \tag{7-40}$$

[1] G. Birkhoff and S. MacLane, "A Survey of Modern Algebra," chap. IX, sec. 2, The Macmillan Company, New York, 1941.

In the last term here we have taken advantage of the evenness of the power spectral densities to take account of the fact that $B_0 = 0$ regardless of the received signal or the transmitted spectrum.

**Problem 7-23.** Find the variances and covariances of the Fourier coefficients for large $T$ by expressing the $A_j$ and $B_j$ directly in terms of $u(t)$. Note that, when $T$ is not large,[1] the $A_j$, because of end effects, are not statistically independent of each other, nor are the $B_j$ independent of each other, and their variances are not equal to each other or to $\Psi(j/T)$. Nevertheless, the representation (7-37) with statistically independent zero-mean normally distributed coefficients having this variance is widely used for ergodic gaussian random processes, and it yields correct results in the limit $T \to \infty$, because the end effects then become negligible.

**Optimum Receiver.** An optimum receiver must evaluate a monotonic function of (7-40). As (7-38) shows, $A_j^2 + B_j^2$ is the energy of $u(t)$ in $(0,T)$ that is attributable to frequencies with absolute value between $(j - \frac{1}{2})/T$ and $(j + \frac{1}{2})/T$. Thus the first summation in (7-40) is a weighted average of the energy spectrum of the received signal $u(t)$. If $\Psi_1(f) \geq \Psi_0(f)$ for all $f$, it can be evaluated (Fig. 7-6$a$) by passing $u(t)$ through a filter whose frequency response has a magnitude $|G(f)|$ proportional to $\sqrt{\dfrac{1}{\Psi_0(f)} - \dfrac{1}{\Psi_1(f)}}$, and integrating the square of the output over the observation interval $(0,T)$ to determine the filter's output energy.

The filter, whose phase shift is immaterial, will take a time of the order of the reciprocal of its bandwidth, which we suppose is small compared to $T$, to settle down to its steady-state response to (7-37). Thereafter it will simply multiply each term by $|G(j/T)|$ and shift its phase, thus yielding (7-40), apart from its constant term, as its total output energy during the interval $(0,T)$.

If $\Psi_1(f) - \Psi_0(f)$ is negative for some frequencies, $u(t)$ may be fed instead (Fig. 7-6$b$) to two filters—one with frequency response

$$|G_1(f)| = 1/\sqrt{\Psi_1(f)}$$

up to a sufficiently high frequency, and the other with

$$|G_0(f)| = 1/\sqrt{\Psi_0(f)},$$

thus whitening the respective input spectra. The integral of the difference of the squares of the filter outputs is then $2L$ except for its constant term.

[1] W. L. Root and T. S. Pitcher, On the Fourier Series Expansion of Random Functions, *Ann. Math. Statist.*, **26**:313–318 (1955); N. M. Blachman, On Fourier Series for Gaussian Noise, *Information and Control*, **1**:56–63 (1957).

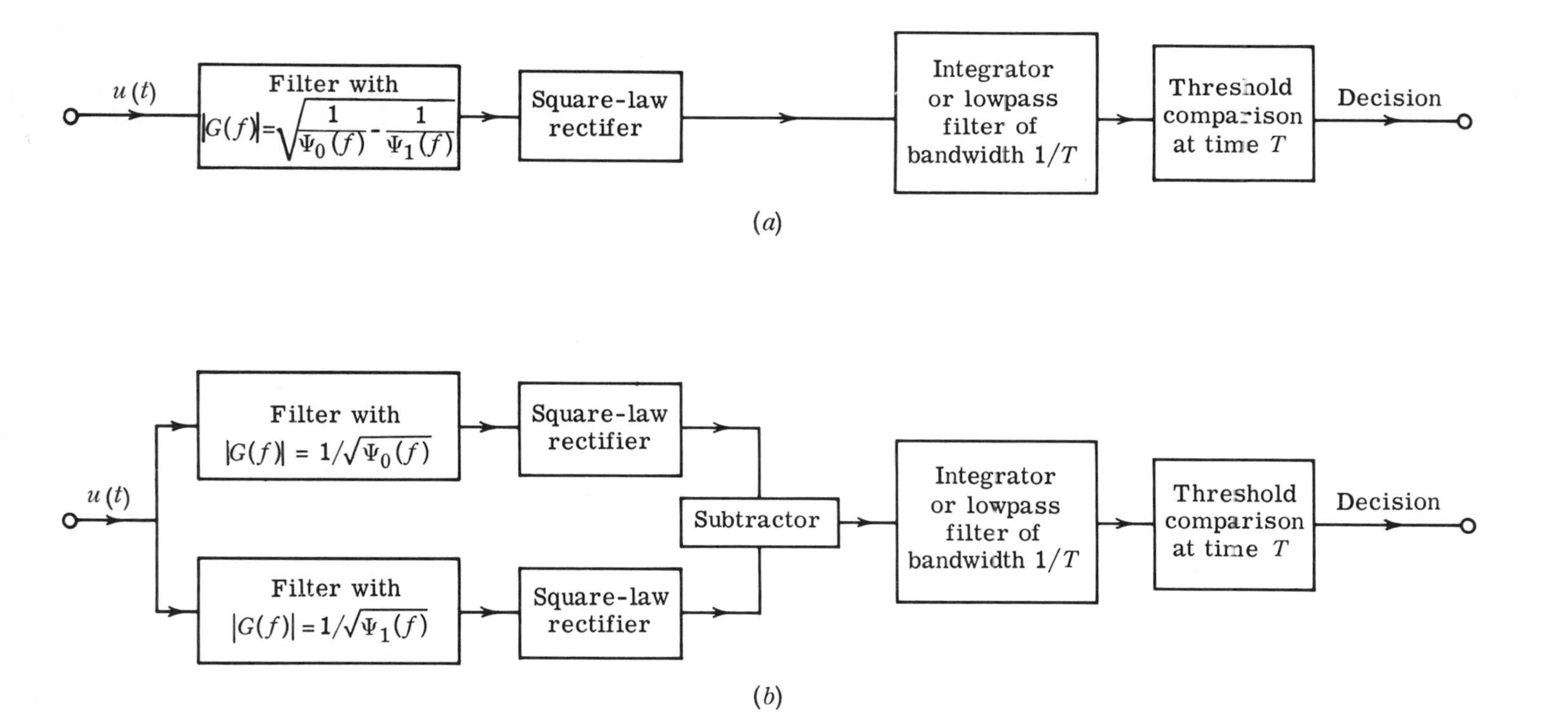


***Figure 7-6.*** Optimum receivers for distinguishing between two gaussian noise-like signals (*a*) in case the spectral density of one never exceeds that of the other and (*b*) in the general case, when the observation time $T$ is long.

**Problem 7-24.** Describe an optimum receiver for discriminating between background noise of power spectral density $\frac{1}{2}N_0$ over all frequencies of interest and the sum of this gaussian background noise plus white gaussian noise of power spectral density $\frac{1}{2}N_1$ from $-W$ to $W$.

**Problem 7-25.** Describe an optimum (diversity) receiver for the case in which two or more statistically independent received signals are available, all representing the same transmitted binary signal, in the cases of coherent, incoherent, and noise-like signals.

**Optimum-Receiver Performance.** When $T$ is large, (7-40) is the sum of a large number of statistically independent terms, whether the transmitted spectrum is $\Psi_0(f)$ or $\Psi_1(f)$. By the central limit theorem, then, its distribution in either case becomes normal, with mean

$$\mathrm{E}\,\{L|0\} = \tfrac{1}{2} \sum_{j=-WT}^{WT} \left[1 - \frac{\Psi_0(j/T)}{\Psi_1(j/T)} - \ln \frac{\Psi_1(j/T)}{\Psi_0(j/T)}\right] \tag{7-41}$$

or

$$\mathrm{E}\,\{L|1\} = \tfrac{1}{2} \sum_{j=-WT}^{WT} \left[\frac{\Psi_1(j/T)}{\Psi_0(j/T)} - 1 - \ln \frac{\Psi_1(j/T)}{\Psi_0(j/T)}\right] \tag{7-42}$$

and variance

$$\begin{aligned} \mathrm{var}\,\{L|0\} &= \tfrac{1}{2} \sum_{j=-WT}^{WT} \left[1 - \frac{\Psi_0(j/T)}{\Psi_1(j/T)}\right]^2 \\ \text{or} \qquad \mathrm{var}\,\{L|1\} &= \tfrac{1}{2} \sum_{j=-WT}^{WT} \left[\frac{\Psi_1(j/T)}{\Psi_0(j/T)} - 1\right]^2. \end{aligned} \tag{7-43}$$

We obtain (7-41) and (7-42) from (7-40) by replacing $A_j^2$ and $B_j^2$ with their mean value and taking advantage of the evenness of the power spectral densities. Similarly, (7-43) is the sum of the variances of the terms of (7-40), which we find by noting that, since $A_j$ and $B_j$ are zero-mean normal random variables, by (1-4) the variance of $A_j^2$ or $B_j^2$ is twice the square of its mean—that is, $\mathrm{E}\,\{A_j^4\} - [\mathrm{E}\,\{A_j^2\}]^2 = 3\sigma^4 - \sigma^4$.

Letting $W$ become infinite now, we see that, for large $T$, (7-41) to (7-43) can be expressed as

$$\mathrm{E}\,\{L|0\} = T \int_0^\infty \left[1 - \frac{\Psi_0(f)}{\Psi_1(f)} - \ln \frac{\Psi_1(f)}{\Psi_0(f)}\right] df = c_0 T, \tag{7-44}$$

$$\mathrm{E}\,\{L|1\} = T \int_0^\infty \left[\frac{\Psi_1(f)}{\Psi_0(f)} - 1 - \ln \frac{\Psi_1(f)}{\Psi_0(f)}\right] df = c_1 T, \tag{7-45}$$

$$\mathrm{var}\,\{L|0\} = T \int_0^\infty \left[1 - \frac{\Psi_0(f)}{\Psi_1(f)}\right]^2 df = d_0^2 T, \tag{7-46}$$

$$\mathrm{var}\,\{L|1\} = T \int_0^\infty \left[\frac{\Psi_1(f)}{\Psi_0(f)} - 1\right]^2 df = d_1^2 T. \tag{7-47}$$

When $L$ exceeds some appropriate threshold, say $\theta$, the inference should be drawn that $\Psi_1(f)$ was the transmitted spectrum. The false-alarm and detection probabilities are then the probability that $L > \theta$ when $\Psi_0(f)$ or $\Psi_1(f)$, respectively, is transmitted, viz., by the central limit theorem,

$$P_F = \operatorname{erf}\left(-\frac{\theta - c_0 T}{d_0 \sqrt{T}}\right) \quad \text{and} \quad P_D = \operatorname{erf}\left(\frac{c_1 T - \theta}{d_1 \sqrt{T}}\right). \tag{7-48}$$

In using (7-48) it should be kept in mind that the sixth power of the argument of each error function must be small compared to $2WT + 1$, the number of random terms in $L$, since the distribution of $L$ will not be normal far out on its tails.

Eliminating $\theta$ between these two equations and solving for $T$, we find that an observation of duration

$$T = \left[\frac{d_1 \operatorname{erf}^{-1}(P_D) - d_0 \operatorname{erf}^{-1}(P_F)}{c_1 - c_0}\right]^2 \tag{7-49}$$

is needed in order to have a detection probability $P_D$ and a false-alarm probability $P_F$. [Note that $\operatorname{erf}^{-1}(P_F)$ is negative for $P_F$ less than $\frac{1}{2}$.] Although we have limited our attention to the case of a predetermined observation time $T$, a "sequential" approach can also be used. There $L$ is calculated after an initial observation and, if it is sufficiently positive or negative, a decision is made in favor of $\Psi_1(f)$ or $\Psi_0(f)$, respectively. Otherwise, the observation is continued until a sufficient weight of evidence has been accumulated to justify a decision one way or the other.

**Problem 7-26.** Because $\ln x$ is a concave function with tangent $x - 1$ at $x = 1$, we have $\ln x \leq x - 1$, with equality only when $x = 1$. Use this inequality to prove that (7-44) is never positive and (7-45) is never negative, the two being equal only when $\Psi_1(f) = \Psi_0(f)$ for all $f$. Also use it to show that, in general, the expected weight of evidence in favor of mark when mark is sent, $\mathrm{E}\{L|1\}$, is not negative and is zero only when $L = \ln \dfrac{p(\mathbf{u}|1)}{p(\mathbf{u}|0)}$ is everywhere zero.

**Problem 7-27.** Show that, in the case of Prob. 7-24, (7-49) becomes $T = [(1 + N_0/N_1) \operatorname{erf}^{-1}(P_D) - (N_0/N_1) \operatorname{erf}^{-1}(P_F)]^2/W$, which is valid if $N_0/N_1 \gg [\operatorname{erf}^{-1}(P_D) - \operatorname{erf}^{-1}(P_F)]^2$. Here the distribution of $L$ is actually "chi-squared with $2WT + 1$ degrees of freedom," which becomes normal for large $T$.

**Problem 7-28.** Show that, when the difference between $\Psi_1(f)$ and $\Psi_0(f)$ is always very small compared to $\Psi_0(f)$, we have $2c_1 = -2c_0 = d_0^2 = d_1^2 = \int_0^\infty [\Psi_1(f) - \Psi_0(f)]^2 \, df/\Psi_0^2(f)$.

**Singular Cases.** If $\Psi_1(f)/\Psi_0(f)$ approaches any limit other than unity as $f \to \infty$, (7-44) will be $-\infty$, and (7-45) will be $+\infty$. That is, an infinite average weight of evidence will be obtained from the observation, regardless of the value of $T$, and an unerring decision can be made as a result of the difference between $\Psi_1(f)$ and $\Psi_0(f)$ at very high frequencies.[1] [The standard deviation of $L$ will also become infinite as the upper limit on the integrals (7-46) and (7-47) increases, but it will be of a smaller order than the expectation of $L$.]

When $\Psi_0(f)$ and $\Psi_1(f)$ are both band-limited, (7-44) to (7-47) are indeterminate. As we have seen, $u(t)$ can be extrapolated indefinitely far in this case, both forward and backward, from any interval of length $T > 0$, and again unerring discrimination is possible. However, if both spectra are augmented by the same small but nonvanishing amount outside their band, the integrands of (7-44) to (7-47) become zero outside the band, and only imperfect discrimination is possible. This is so because the additional high-frequency noise limits the extent to which $u(t)$ can be accurately extrapolated. Although weak additional noise permits extrapolation over a long interval, the latter is negligible by comparison with $T$, which we have assumed to be large.

Such noise will inevitably be present in the input to any receiver because of the physical nature of the apparatus with its constituent electrons and because of the impossibility of measuring any value of $u(t)$ with infinite accuracy. These effects contribute to $\Psi_0(f)$ and $\Psi_1(f)$ a background noise that prevents them from vanishing at any frequency and dominates both of them as $f \to \infty$, so that their ratio approaches unity, and perfectly reliable or "singular" detection is not possible.[2]

Even though $\Psi_1(f)/\Psi_0(f) \to 1$ as $f$ grows infinite, (7-44) and (7-45) may be $-\infty$ and $+\infty$, respectively. The result of Prob. 7-28 shows that, as the spectra approach forms for which this is the case, $\mathrm{E}\{L|1\} - \mathrm{E}\{L|0\}$ will be of a larger order than the standard deviations of the two distributions of $L$, and the discrimination between the two spectra becomes singular. This cannot occur, however, if $\Psi_0(f)$ approaches a constant as $f \to \infty$, if $\Psi_1(f)$ is $\Psi_0(f)$ plus a nonnegative increment, and if the integral of this increment is finite, representing a finite total power.

On the other hand, as we see again from Prob. 7-28, a difference $\Psi_1(f) - \Psi_0(f)$ that is, for example, of the order of $\Psi_0(f)/f$ for large $f$ will not result in singular detection. In such a case (7-44) and (7-45) become $+\infty$ if the logarithmic terms are omitted, and hence the first summation in (7-40), which represents the input to the threshold comparison in Fig.

[1] David Slepian, Some Comments on the Detection of Gaussian Signals in Gaussian Noise, *IRE Trans. Inform. Theory,* **IT-4**:65–68 (1958).

[2] Wainstein and Zubakov, *op. cit.,* app. III.

7-6*a*, will certainly be infinite, along with the required threshold. From the nonsingularity, it follows that the highest-frequency components of $u(t)$ carry a vanishing weight of evidence, and so the receiver's performance will be degraded arbitrarily little if the input filter's response is cut off at a sufficiently high frequency. As a result, all quantities become finite, and optimum detection can be approached though not attained.

Since (7-46) and (7-47) can vanish only when $\Psi_0(f)$ and $\Psi_1(f)$ are identical, singular detection must always involve an infinite value[1] for $E\{L|1\} - E\{L|0\}$ (together with standard deviations which, though infinite, are of a smaller order of magnitude as the spectra approach their limiting forms). It cannot result from the vanishing of $\Psi_0(f)$ or $\Psi_1(f)$ over a finite band of frequencies, however, because only a finite number of Fourier coefficients are involved, and the end effects due to a finite observation time will always introduce the missing frequencies. Although this conclusion is confirmed by (7-44) to (7-47), which yield a standard deviation of the same order as the difference of the two means in this case, these equations are not reliable here, because they do not include the end effects, which are particularly important when one of the power spectral densities vanishes. To include these end effects properly, we would have to use the previously mentioned much more involved approach that is applicable to intermediate as well as to short and to long observation times. (See the footnote on p. 140.)

The results of Probs. 7-11 and 7-16 show that discrimination between two *known* signals can also be singular, viz., when the integral appearing in Prob. 7-11 is infinite. The observation time is there assumed infinite, but there are also cases of finite observation time in which known signals are perfectly distinguishable, e.g., when the signals have different discontinuities which the noise cannot have. Any case that is singular for a finite observation time, of course, is also singular for an infinite observation time. However, singular detection will never arise in practical applications, because discontinuous signals are not realizable.

In this chapter we have dealt only with the problem of distinguishing among a discrete set of possible signals in the presence of noise. If, instead, the signals are indexed by a continuous parameter, the reception problem is described as "parameter estimation"[2] rather than detection

[1] See Theorem 2 of D. Middleton, A Note on Singular and Nonsingular Optimum (Bayes) Tests for the Detection of Normal Stochastic Signals in Normal Noise, *IRE Trans. Inform. Theory*, **IT-7**:105–113 (1961); D. Middleton, Remarks on and Revisions of Some Earlier Results in Two Recent Papers, *IRE Trans. Inform. Theory*, **IT-8**:385–387 (1962).

[2] D. Middleton, "An Introduction to Statistical Communication Theory," chap. 21, McGraw-Hill Book Company, New York, 1960, and "Topics in Communication Theory," chap. 3, McGraw-Hill Book Company, New York, 1965.

and recognition. The signal-to-noise ratios we found in Chaps. 5 and 6 measured the accuracy of this parameter (modulating-waveform) estimation. However, they describe only one aspect of the effect of the noise and take no account of the predictability or unpredictability of the variations of the parameter with time. Such problems can be treated by the application of C. E. Shannon's measure of information, which we shall take up in Chap. 8.

Here we have been concerned mainly with the optimum reception of given signals in the presence of noise. The problem of choosing the best possible signals, subject to suitable restrictions, for use along with an optimum receiver is generally more difficult. In Chaps. 8 and 9, however, we shall obtain some solutions to it by means of information theory, although other criteria of optimality may often be more appropriate than information measures.

# PART III

# Information Theory

# chapter 8 INFORMATION THEORY IN THE DISCRETE CASE

In Chap. 7 we used the notion of *information* qualitatively, observing that some received-signal parameters $y$ gave no information about the transmitted signal $x$ because of their statistical independence of it. In general, we may define *information about something*, say $x$, as anything, say $y$, that alters our probability distribution for $x$, making the posterior probabilities $P(x|y)$ different from the prior probabilities $P(x)$ of the various possible values of $x$. These values need not be numerical but can be "mark" or "space," a letter, a sequence of symbols, a signal, a message, and so on, or several of these things taken together, which may come from different sources. For example, (7-1) shows the effect of knowing the received signal (or the likelihood ratio derived from it) upon the probabilities of the two possible transmitted signals.

In general, the probabilities $P(x)$ or $P(x|y)$ describe our (prior or posterior) uncertainty as to the actual value of $x$, and we shall define the amount of information $y$ gives us about $x$ as the amount by which it reduces our uncertainty about $x$, that is, the difference between our prior and posterior uncertainties. Hence we need to find a suitable measure of uncertainty.

## ☐ ENTROPY—THE MEASURE OF UNCERTAINTY

We shall denote the measures of uncertainty associated with the prior probabilities $P(x)$ and the posterior probabilities $P(x|y)$ by $H(X)$ and $H(X|y)$, respectively. Here $X$ is not an argument, but rather, a label indicating the uncertainty about $x$; however, $H(X|y)$ *does* depend on $y$. We shall call $H(X)$ and $H(X|y)$ *entropies*, in accord with the usage of this term in statistical thermodynamics, $H(X|y)$ being a *conditional entropy*. When averaged over the distribution of $y$, it becomes the *average conditional entropy*

$$H(X|Y) = \sum_j P(y_j)H(X|y_j)$$

or, for short,

$$H(X|Y) = \sum_y P(y)H(X|y), \tag{8-1}$$

where the summation is over all possible values $y_j$ of $y$. Thus $H(X|Y)$ is not a function of $x$ or $y$; the label $Y$ simply indicates its relationship to the random variable $y$.

If we let $z$ denote the pair $(x,y)$, we can denote its entropy by either $H(Z)$ or $H(X,Y)$. The latter is called a *joint entropy*. We want all these entropies to take the same form, differing only as to the probability distribution whose uncertainty they express. Thus we suppose that each is a continuous function of the probabilities in this distribution—$P(x)$ in the case of $H(X)$.

In order that these entropies accord with our intuitive feelings about uncertainty, we shall require them to have two further properties.[1] First, we suppose that

$$H(X,Y) = H(X) + H(Y|X), \tag{8-2}$$

that is, that our uncertainty about the pair of random variables $x$ and $y$ equals the sum of our uncertainty about $x$ plus our average uncertainty about $y$ when we already know $x$.

Second, we suppose that, when $x$ takes each of $n$ values with the same probability $1/n$, $H(X)$ is an increasing function of $n$, say $h(n)$; we make this assumption because the larger $n$ is, the greater is our uncertainty about the value of $x$. If $n = 1$ and $x$ is confined to a single value, there is no uncertainty about $x$, and we therefore expect that $h(1) = 0$.

We can clarify the foregoing by reference to a roulette wheel, whose circumference is divided into 38 parts numbered 00, 0, 1, 2, . . . , 36 (in Europe there is no 00), 00 and 0 being green, 18 numbers red, and the remaining 18 black. After the wheel is spun, a ball lands on one of these, say $z$, all 38 being equally likely. The outcome $z$ can be regarded as resulting from a selection of the color, say $x$—with probabilities 18/38 for red and for black and 2/38 for green—followed by the choice of one of the 2 or 18 numbers $y$ having this color—both or all 18 with equal probability. Here our uncertainty as to the overall outcome $z = (x,y)$ is $H(Z) = H(X,Y) = h(38)$. When we know that the color is red or black, our uncertainty as to the number is only $H(Y|\text{red}) = H(Y|\text{black}) = h(18)$.

[1] C. E. Shannon, A Mathematical Theory of Communication, *Bell System Tech. J.*, **27**:379–423 and 623–656 (1948), app. 2, pp. 419–420; reprinted in C. E. Shannon and Warren Weaver, "The Mathematical Theory of Communication," University of Illinois Press, Urbana, Ill., 1949.

If the color is green, our uncertainty as to the number is still smaller, namely, $H(Y|\text{green}) = h(2)$. Hence, by (8-1), the average conditional entropy of the number $y$, given the color $x$, is

$$H(Y|X) = \tfrac{36}{38}h(18) + \tfrac{2}{38}h(2),$$

and we have, from (8-2),

$$h(38) = H(X) + \tfrac{36}{38}h(18) + \tfrac{2}{38}h(2). \tag{8-3}$$

To avoid confusion between $y$ and $z$ here, we may take $y$ to be an ordinal number between first and eighteenth specifying which of the numbers of the color $x$ resulted from the spin without naming the number itself.

**Equal Probabilities.** Evidently, we shall be able to determine $H(X)$ if we can find what $h(n)$, if any, fulfills the foregoing assumptions. If $y$ is statistically independent of $x$, (8-2) becomes

$$H(X,Y) = H(X) + H(Y),$$

and if $x$ takes $m$ different values with equal probability and $y$ independently takes $n$ different values with equal probability, then $(x,y)$ takes $mn$ different values with equal probability, and this equation becomes

$$h(mn) = h(m) + h(n).$$

Hence $h(m^2) = 2h(m)$, and more generally, for any positive integers, $m$ and $n$, and nonnegative integers, $r$ and $s$, we have

$$h(m^r) = rh(m) \qquad \text{and} \qquad h(n^s) = sh(n). \tag{8-4}$$

For any $n$ greater than 1, $m^r$ either will lie between a pair of successive powers of $n$ or will be equal to one of them; i.e., there is exactly one $s$ such that

$$n^s \leq m^r < n^{s+1} \tag{8-5}$$

or

$$s \log_b n \leq r \log_b m < (s + 1) \log_b n \tag{8-6}$$

for any base $b$ greater than 1.

Because of the monotonicity of $h(\cdot)$, we have, from (8-4) and (8-5),

$$sh(n) \leq rh(m) < (s + 1)h(n).$$

Dividing by (8-6), we get

$$\frac{s}{s+1}\frac{h(n)}{\log_b n} < \frac{h(m)}{\log_b m} < \frac{s+1}{s}\frac{h(n)}{\log_b n}.$$

Since these inequalities must hold for any $r$, however large, and hence for arbitrarily large $s$, we must have

$$\frac{h(n)}{\log_b n} = \frac{h(m)}{\log_b m},$$

that is, $h(n)$ must be proportional to $\log_b n$. Since any change in $b$ multiplies $\log_b n$ by a constant, we can absorb the constant of proportionality into the base, getting

$$h(n) = \log_b n. \tag{8-7}$$

To fulfill our condition that $h(n)$ be an increasing function of $n$, it is only necessary that the base $b$ exceed 1; otherwise, it is arbitrary.

If $b = 2$, we say that the entropy is measured in *binary units* or *bits*. If the natural logarithm is used, $h(n)$ will be smaller by the factor $\ln 2 = 0.693$, and we associate the *natural unit* or *nit* with the resulting value, 1 nit being 1.44 bits. The bit is the amount of uncertainty, $h(2)$, as to the result of flipping a fair coin. The nit has no such simple interpretation, but natural logarithms are sometimes convenient. Common logarithms can also be used; they measure entropy in *decimal units* or *dits*. Henceforth all our logarithms will be understood to have the base $b$ unless another base is indicated.

**Problem 8-1.** How many bits make one dit? How many nits?

**Problem 8-2.** Using (8-3) and (8-7), find $H(X)$, the entropy of the roulette-wheel color, in bits.

**Unequal Probabilities.** When $x$ takes a finite set of $n$ values with *unequal* probabilities, we can use (8-2) to find its entropy $H(X)$ by expressing or approximating these probabilities as the quotients of integers, namely, $P(x_1) = m_1/N$, $P(x_2) = m_2/N$, . . . , $P(x_n) = m_n/N$, where $N = m_1 + \cdots + m_n$. We now let $y$ be a random variable which, when $x$ is $x_i$, takes $m_i$ different values with equal probability. If $x$ is the roulette-wheel color, $m_1$ and $m_2$ may be 18 (or 9), and $m_3$ may be 2 (or 1) for $x_1$ = red, $x_2$ = black, and $x_3$ = green.

As a result, the pair $(x,y)$ takes $N$ different values with equal probability, and $H(X,Y) = \log N$. The conditional entropy of $y$, given that $x$ is $x_i$, is $H(Y|x_i) = \log m_i$, by (8-7), and its average conditional entropy, by (8-1), is thus

$$H(Y|X) = \sum_{i=1}^{n} \frac{m_i}{N} \log m_i.$$

Subtracting this from $H(X,Y) = \log N$, we get, by (8-2),

$$\begin{aligned} H(X) &= H(X,Y) - H(Y|X) \\ &= \log N - \sum_{i=1}^{n} \frac{m_i}{N} \log m_i \\ &= - \sum_{i=1}^{n} \frac{m_i}{N} \log \frac{m_i}{N} \\ &= - \sum_{i=1}^{n} P(x_i) \log P(x_i) \end{aligned}$$

or, for short,

$$H(X) = - \sum_{x} P(x) \log P(x). \tag{8-8}$$

We shall use (8-8) as our measure of uncertainty, whether or not the $P(x_i)$ are rational and can be expressed exactly as $m_i/N$, since $H(X)$ was to be a continuous function of the $P(x_i)$. We shall use it when $n$ is infinite (see Prob. 8-3) as well as when it is finite, but we can apply it only to discrete distributions. In the continuous case, each $P(x)$ has the form $p(x)\,dx$, and the $dx$ in the argument of the logarithm makes (8-8) infinite, for our uncertainty as to the exact value of $x$—down to the "last" decimal place—really is infinite. We shall therefore confine our attention in this chapter to discrete distributions, postponing until Chap. 9 the modifications that are necessary in the continuous case.

When any $P(x)$ is zero, we omit the corresponding term from (8-8), since such an $x$ will almost never occur, and furthermore, the limit of $P \log P$ as $P \to 0$ is zero. Since $-P \log P \geq 0$ for $0 < P \leq 1$, (8-8) is never negative, and it attains its least possible value, zero, only when every term vanishes, i.e., when every $P(x)$ is either 0 or 1. Thus $H(X) = 0$ if and only if one $x_i$ has probability 1 and the rest have probability 0, that is, when there is no uncertainty at all about the value of $x$.

On the other hand, making use of the inequality (see Prob. 7-26)

$$\log u \leq (u - 1) \log e, \tag{8-9}$$

which becomes an equality only at $u = 1$, where the right-hand side is the tangent to the left-hand side (Fig. 8-1), we find that

$$\begin{aligned} H(X) &= \sum_{i=1}^{n} P(x_i) \log \frac{1}{nP(x_i)} + \log n \\ &\leq \sum_{i=1}^{n} \left[ \frac{1}{n} - P(x_i) \right] \log e + \log n \\ &= \log n, \end{aligned} \tag{8-10}$$

since

$$\sum_{i=1}^{n} P(x_i) = 1. \tag{8-11}$$

Thus, for a fixed number $n$ of different values of $x$, $H(X)$ never exceeds $\log n$, and it attains this maximum value if and only if all $n$ values have the same probability, $1/n$.

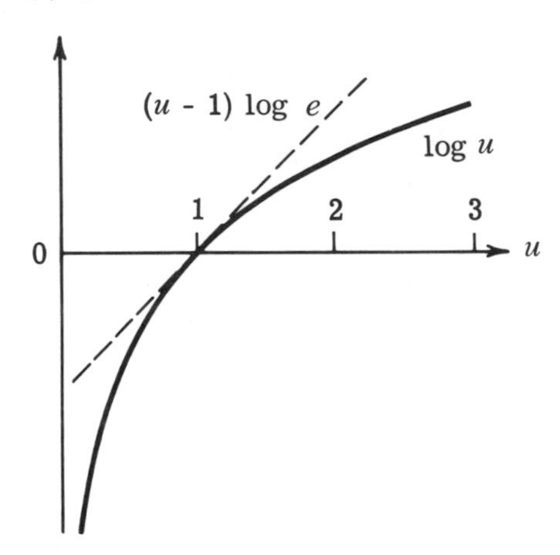


*Figure 8-1.* The function $\log u$ and an upper bound for it.

**Problem 8-3.** A fair coin is tossed repeatedly until it comes up tails. Show that the entropy of the number of tosses in this experiment is 2 bits.

**Problem 8-4.** What is the entropy of the result of rolling a pair of dice of different colors? What would it be if the dice were indistinguishable? What would it be in each case if it were known that at least one of the dice had come up six? What is the entropy of the total number of spots (between 2 and 12)? What is the entropy of the difference (between $-5$ and 5) between the dice?

**Problem 8-5.** Draw a graph of the entropy of the toss of a biased coin as a function of the probability $P$ of heads, noticing that the curve is symmetric about its maximum ($P = \frac{1}{2}$) and has an infinite slope at both ends ($P = 0$ and $P = 1$).

**Problem 8-6.** Find the entropy of the discrete distribution with probabilities $P_j = c/(j \log^2 j)$ $(j = 2, 3, \ldots)$, with $c$ chosen to make $P_2 + P_3 + \cdots = 1$.

**Problem 8-7.** A fair coin is tossed repeatedly until it has come up heads $N$ times altogether. What is the entropy of the resulting sequence of heads and tails?

## INFORMATION

The amount of our (prior) uncertainty as to $x$ is $H(X)$, given by (8-8). If we find out the actual value of $x$, the entropy of the posterior distribution of $x$ will be $H(X|y) = 0$, that is, the information $y$ reduces our uncertainty about $x$ to zero. The amount of this reduction, $H(X)$, is a measure of the information provided by $y$. Note that this measure will be the same regardless of which value $x$ turns out to have, for our posterior uncertainty in *any* case is zero.[1]

In general, however, the information $y$ will not reveal the value of $x$ with complete certainty but will yield a set of posterior probabilities $P(x|y)$ and a posterior entropy or uncertainty about $x$,

$$H(X|y) = -\sum_x P(x|y) \log P(x|y), \tag{8-12}$$

given by (8-8) with $P(x)$ replaced by $P(x|y)$. Like (8-8), (8-12) cannot be negative. Notice that (8-8) can be used to evaluate *any* entropy, provided that $P(x)$ is replaced by the probabilities of a set of mutually exclusive events whose total probability is 1.

The amount by which $y$ reduces our uncertainty as to $x$ is then our measure of the amount of information it gives about $x$,

$$I(X;y) = H(X) - H(X|y). \tag{8-13}$$

This quantity can be positive, e.g., when $y$ tells us just which value $x$ has. It can be zero, e.g., when $y$ is completely irrelevant to $x$ and we have $P(x|y) = P(x)$ and $H(X|y) = H(X)$. And it can be negative, as when $y$ increases our uncertainty—a phenomenon that often results from reading a newspaper (see Prob. 8-8 below).

The *average* amount of information that $y$ gives about $x$ is the average of (8-13) over the distribution of $y$,

$$I(X;Y) = \sum_j P(y_j)I(X;y_j)$$

or, for short,

$$I(X;Y) = \sum_y P(y)I(X;y).$$

[1] Many authors take a different view, attributing to an outcome of prior probability $P$ an amount of information $-\log P$. In either case, the *average* amount of information is the same.

Thus, making use of (8-1) and (8-2), we have[1]

$$\begin{aligned} I(X;Y) &= H(X) - H(X|Y) \\ &= H(X) + H(Y) - H(X,Y) \\ &= H(Y) - H(Y|X). \end{aligned} \tag{8-14}$$

Because of its symmetry, (8-14) is said to measure the *mutual information*[2] of $x$ and $y$; here again $X$ and $Y$ serve as labels rather than as arguments. The three equalities (8-14) can be represented graphically as in Fig. 8-2.

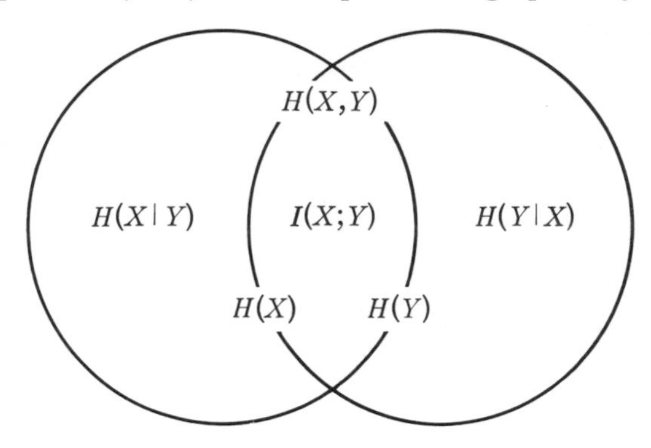


*Figure 8-2.* A graphical representation of various relationships among entropies and mutual information. Each quantity is represented by the area in which its symbol appears, for example, $I(X;Y)$ by the lenticular area in the center and $H(X,Y)$ by the entire figure.

Making use of (8-8) and (8-12) and the equalities

$$P(x,y) = P(x)P(y|x) = P(y)P(x|y),$$

or noticing that (8-8) gives us

$$H(X,Y) = -\sum_{x,y} P(x,y) \log P(x,y), \tag{8-15}$$

we find

$$\begin{aligned} I(X;Y) &= \sum_{x,y} P(x,y) \log \frac{P(x|y)}{P(x)} \\ &= \sum_{x,y} P(x,y) \log \frac{P(y|x)}{P(y)} \\ &= \sum_{x,y} P(x,y) \log \frac{P(x,y)}{P(x)P(y)}. \end{aligned} \tag{8-16}$$

[1] Many authors define $I(X; y)$ differently from (8-13), but the average value of their information measure is nevertheless (8-14). See Prob. 9-8.

[2] Because $I(X;X) = H(X)$, this quantity is sometimes called the *self-information* of $x$.

Like (8-8), (8-16) can never be negative, for, by (8-9),

$$I(X;Y) = -\sum_{x,y} P(x)P(y|x) \log \frac{P(y)}{P(y|x)}$$

$$\geq -\sum_{x,y} P(x)P(y|x) \left[\frac{P(y)}{P(y|x)} - 1\right] \log e$$

$$= 0, \tag{8-17}$$

with equality if and only if $P(y) = P(y|x)$ for every $y$ and for every $x$ with $P(x) \neq 0$, that is, when $y$ is statistically independent of $x$. Thus, whenever $x$ and $y$ are not statistically independent, $y$ will, on the average, give a positive amount of information about $x$.

From (8-14) and (8-17), we see that

$$H(X|Y) \leq H(X); \tag{8-18}$$

that is, on the average, conditioning never increases an entropy.

**Problem 8-8.** How much information does the roulette-wheel color $x$ give about the number $z$ (between 00 and 36) when $x$ = red? Green? On the average? Find $I(X;y)$, where $y$ is the ordinal number (between first and eighteenth) of $z$ among the numbers having the color $x$. Notice that it is sometimes negative, and find $I(X;Y)$.

**Problem 8-9.** Using the results of Prob. 8-4, determine how much information, on the average, the total of two dice gives about the number of spots showing on each. Note that it is sometimes more convenient to use the first line of (8-14) and sometimes the last.

**Problem 8-10.** A *symmetric binary channel* has an input $x$ and an output $y$ which take only the values 0 and 1, $y$ being the same as $x$ with probability $P$. Determine all of the quantities shown in Fig. 8-2 for the case in which $x$ takes each value with probability $\frac{1}{2}$.

**Problem 8-11.** In the symmetric *binary erasure channel* the input $x$ is either 0 or 1, and the output $y$ is the same as $x$ with probability $P$, is the other binary symbol with probability $Q$, and is a third symbol (erasure) with probability $1 - P - Q$. Determine all of the quantities shown in Fig. 8-2 for the case in which $x$ takes each value with probability $\frac{1}{2}$.

**Conditional Information.** If, already knowing $y$, we obtain additional information $z$, our uncertainty about $x$ falls from $H(X|y)$ to $H(X|y,z)$. This decrease,

$$I(X;z|y) = H(X|y) - H(X|y,z), \tag{8-19}$$

measures the amount of this additional information about $x$. It may be called the conditional information that $z$ gives about $x$ for a given $y$, and it is of the same form as (8-13), with $y$ replaced by $z$ and each term conditioned on $y$.

When added to (8-13), (8-19) becomes

$$\begin{aligned} I(X;y) + I(X;z|y) &= H(X) - H(X|y,z) \\ &= I(X;y,z), \end{aligned} \tag{8-20}$$

which is the total amount of information that $y$ and $z$ together give about $x$. Averaging (8-19) over all $y$ and $z$, we get

$$\begin{aligned} I(X;Z|Y) &= H(X|Y) - H(X|Y,Z) \\ &= I(X;Y,Z) - I(X;Y). \end{aligned} \tag{8-21}$$

Like (8-17), (8-21) cannot be negative. However, because it can be positive even when $I(X;Z) = 0$ (see Prob. 8-13), a diagram for it[1] like Fig. 8-2 can be misleading.

**Problem 8-12.** Find $I(X;Z|Y)$ with $x$ the result of rolling a pair of distinguishable dice, $y$ the total number of spots, and $z$ the signed difference.

**Problem 8-13.** Find $I(X;Y)$, $I(X;Z|Y)$ and $I(X;Y,Z)$ when $(x,y,z)$ takes the four values (0,0,0), (0,1,1), (1,0,1), (1,1,0), each with probability $\frac{1}{4}$.

**Problem 8-14.** Show that the distribution over the nonnegative integers with mean $\mu$ having the maximum entropy is the geometric distribution, and use (8-9) to show that its entropy, $\log(\mu + 1) + \mu \log(\mu + 1)/\mu$, does not exceed $\log(\mu + 1)e$.

## ☐ ENTROPY AND INFORMATION RATES

We shall now consider the entropy of an ergodic sequence of random symbols . . . , $x_{-1}$, $x_0$, $x_1$, $x_2$, . . . , possibly spelling out messages, which may be regarded as the output of an ergodic "information source." (Here the $x_i$ are the successive output symbols rather than the possible values of $x$ as previously.) Because of the ergodicity, the entropy of every set of $k$ successive symbols is the same as that of $x_1, x_2, \ldots, x_k$, which, by (8-2), can be expressed as

$$\begin{aligned} H(X_1, \ldots, X_k) = H(X_1) + H(X_2|X_1) + \cdots \\ + H(X_k|X_{k-1}, \ldots, X_1). \end{aligned} \tag{8-22}$$

[1] N. M. Blachman, A Generalization of "Mutual Information," *Proc. IRE*, **49**:1331–1332 (1961).

By (8-18) we see that the terms on the right-hand side form a nonincreasing sequence, bounded below by zero, since no entropy can be negative. Hence the limit of $H(X_k|X_{k-1},X_{k-2}, \ldots ,X_{k-j})$ as $j \to \infty$ must exist, though its value can be infinite. We shall call this limit the *entropy rate per symbol* for reasons which will shortly become clear, and we shall denote it by

$$H'(\mathbf{X})/\nu = -\lim_{j\to\infty} \sum_{x_{k-j}, \ldots ,x_k} P(x_{k-j}, \ldots ,x_k) \log P(x_k|x_{k-1}, \ldots ,x_{k-j})$$
$$= H(X_k|X_{k-1},X_{k-2}, \ldots), \tag{8-23}$$

where $\nu$ is the average number of symbols per unit time and $H'(\mathbf{X})$ the *entropy rate per unit time.*

As $k \to \infty$, all of the terms of (8-22) except a vanishing fraction at the beginning approach (8-23), and we see that

$$\lim_{k\to\infty} \frac{H(X_1, \ldots ,X_k)}{k} = \frac{H'(\mathbf{X})}{\nu}.$$

Hence, for large $k$,

$$H(X_1, \ldots ,X_k) = H'(\mathbf{X})(k/\nu) + o(k), \tag{8-24}$$

where $o(k)$ is a quantity of smaller order than $k$, i.e., such that $o(k)/k \to 0$ as $k \to \infty$.

If, like dot and dash, the various possible symbols do not have equal durations and they delay the succeeding symbol by different amounts, the number $k$ of output symbols $x_i$ occurring during any given time interval will be a random variable. The resulting uncertainty as to the output during an interval of length $T$ is, by (8-1), (8-2), and (8-8),

$$H(K,X_1, \ldots ,X_K) = -\sum_{x_1, \ldots ,x_k,k} P(k,x_1, \ldots ,x_k) \log P(k,x_1, \ldots ,x_k)$$
$$= H(K) + \sum_{k=0}^{\infty} P(k)H(X_1, \ldots ,X_k). \tag{8-25}$$

From the result of Prob. 8-14 we see that, since $\nu T$ is the expected number of symbols in time $T$, $H(K)$ cannot exceed $\log (\nu T + 1)e$. Hence it disappears when we divide (8-25) by $T$ and let $T \to \infty$. Making use of (8-24), we thus find that, inasmuch as $\Sigma_k P(k)k/\nu = T$,

$$\lim_{T\to\infty} \frac{H(K,X_1, \ldots ,X_K)}{T} = \lim_{T\to\infty} \frac{H'(\mathbf{X})T + o(\nu T)}{T}$$
$$= H'(\mathbf{X}). \tag{8-26}$$

Since

$$P(k,x_1, \ldots ,x_k) = P(k)P(x_1)P(x_2|x_1) \cdots P(x_k|x_{k-1}, \ldots ,x_1),$$

we have

$$-\frac{\log P(k,x_1, \ldots ,x_k)}{T} = -\frac{\log P(k)}{T} + \frac{k}{T}\,\frac{-\log P(x_1) - \log P(x_2|x_1) - \cdots - \log P(x_k|x_{k-1}, \ldots , x_1)}{k}. \tag{8-27}$$

Since the first term on the right cannot be negative, and since its expectation, $H(K)/T$, is at most of the order of $T^{-1} \log \nu T$, it vanishes in the limit $T \to \infty$. In the same limit, the quotient $k/T$ approaches $\nu$, and the fraction it multiplies evidently becomes a time average of

$$-\log P(x_k|x_{k-1},x_{k-2}, \ldots),$$

whose ensemble average, by (8-23), is $H'(\mathbf{X})/\nu$. Thus, since the sequence of $x_i$ is ergodic, the only limit that (8-27) can approach is $H'(\mathbf{X})$, in agreement with (8-26).

By the law of large numbers,[1] therefore, (8-27) for large $T$ is very likely to be close to $H'(\mathbf{X})$. More precisely, for any positive $\delta$ and $\epsilon$, for sufficiently large $T$, the probability that

$$H'(\mathbf{X}) - \epsilon \leq -\frac{1}{T} \log P(k,x_1, \ldots ,x_k) \leq H'(\mathbf{X}) + \epsilon \tag{8-28}$$

will be at least $1 - \delta$.

**The Sets of Typical and Atypical Sequences.** We shall use the inequalities (8-28) to separate the possible information-source output sequences during a sufficiently long time $T$ into two sets. We shall call those for which

$$b^{-[H'(\mathbf{X})+\epsilon]T} \leq P(k,x_1, \ldots ,x_k) \leq b^{-[H'(\mathbf{X})-\epsilon]T} \tag{8-29}$$

*typical* and the rest *atypical*, $b$ being the base of our logarithms. The number of typical sequences must lie between

$$(1 - \delta)b^{[H'(\mathbf{X})-\epsilon]T} \qquad \text{and} \qquad b^{[H'(\mathbf{X})+\epsilon]T}, \tag{8-30}$$

since at least the former number with the largest probability satisfying (8-29) would be needed to yield a total probability $1 - \delta$, and there

[1] William Feller, "Probability Theory and Its Applications," vol. 1, chap. 10, p. 191, John Wiley & Sons, Inc., New York, 1950. The law of large numbers is applicable to a very large class of ergodic information sources; see *ibid.*, vol. 2, p. 240, 1966. We shall leave aside questions as to the applicability of this (weak) law of large numbers to specific information sources in order to avoid restricting and complicating the discussion. Rigorous derivations for specific cases of the results we thus obtain heuristically can be found, for example, in M. S. Pinsker, "Information and Information Stability of Random Variables and Processes," Amiel Feinstein (trans.), Holden-Day, Inc., San Francisco, 1964; Robert B. Ash, "Information Theory," Interscience Publishers, Inc., New York, 1965; and other works cited further on.

could be no more than the latter number with the smallest probability satisfying (8-29).

**Information Rate.** Given two jointly ergodic random sequences $\ldots, x_{-1}, x_0, x_1, \ldots$ and $\ldots, y_{-1}, y_0, y_1, \ldots$, we can generalize the definition (8-26) of entropy rate to give the joint entropy rate $H'(\mathbf{X},\mathbf{Y})$ of the two sequences and the conditional entropy rates $H'(\mathbf{X}|\mathbf{Y})$ and $H'(\mathbf{Y}|\mathbf{X})$. Notice that $H'(\mathbf{X}|\mathbf{Y})$, for example, is the average entropy rate of the $x$ sequence, given the *entire* $y$ sequence.

As $T$ grows, our uncertainty about the $x$ sequence during this time grows at the rate $H'(\mathbf{X})$, by (8-26). However, if we are able to find out the $y$ sequence, our uncertainty will grow only at the rate $H'(\mathbf{X}|\mathbf{Y})$. The difference between these two rates,

$$\begin{aligned} I'(\mathbf{X};\mathbf{Y}) &= H'(\mathbf{X}) - H'(\mathbf{X}|\mathbf{Y}) \\ &= H'(\mathbf{X}) + H'(\mathbf{Y}) - H'(\mathbf{X},\mathbf{Y}) \\ &= H'(\mathbf{Y}) - H'(\mathbf{Y}|\mathbf{X}), \end{aligned} \tag{8-31}$$

then, is the rate at which the $y$ sequence conveys information about the $x$ sequence, i.e., the mutual-information rate.

Entropy and information rates can be measured in bits, nits, or dits per second. If such a value is divided by the average number of symbols per second, the result will be the rate in bits, nits, or dits per symbol. If divided by the average cost per unit time, we obtain the rate per unit cost.[1] The cost may represent the electrical energy needed to transmit the symbols, possibly different for each symbol in the alphabet of the information source.

**Problem 8-15.** Show that, if the $x$ sequence and the $y$ sequence both have $\nu$ symbols per second on the average, and if each $y_i$ is statistically dependent only on $x_i$, being independent of all other symbols in either sequence, then $H'(\mathbf{Y}) = \nu H(Y_k)$, $H'(\mathbf{Y}|\mathbf{X}) = \nu H(Y_k|X_k)$, and $I'(\mathbf{X};\mathbf{Y}) = \nu I(X_k;Y_k)$, which is of the form (8-16). What are the mutual information rates for the symmetric binary and binary erasure channels (Probs. 8-10 and 8-11) if they can be used $\nu$ times per second and each time the input and output are statistically independent of their predecessors?

## □ THE FUNDAMENTAL CODING THEOREM FOR A DISCRETE INFORMATION SOURCE

We shall now consider the $b$-ary coding of the output of our information source; i.e., we shall represent each possible output sequence, say $x_1, \ldots, x_k$ during a long time $T$ by a sequence of symbols chosen from

[1] N. M. Blachman, Minimum-Cost Encoding of Information, *IRE Trans. Inform. Theory*, **PGIT-3**:139–149 (1954); N. M. Blachman, Minimum-Cost Transmission of Information, *Information and Control*, **7**:508–511 (1964).

an alphabet of size $b$, such as the digits 0, 1, 2, . . . , $b - 1$, where $b$ is an integer greater than 1. Using code words of length $L$ $b$-ary digits, we have $b^L$ code words available, from 00 . . . 0 to $(b - 1)(b - 1) \ldots (b - 1)$, and these will suffice to provide a different code word for each $x$ sequence or "message" in the typical set and one code word to be used for *all* atypical messages if $b^L$ is at least the upper bound of (8-30), that is, if $L \geq [H'(\mathbf{X}) + \epsilon]T$. Dividing by $T$, we see that any code-digit rate $L/T$ greater than the entropy rate $H'(\mathbf{X})$ will suffice to permit coding which can be decoded correctly with any desired probability $1 - \delta$ that is less than 1. This is the first half of the fundamental coding theorem for a discrete information source.

To see if a *smaller* code-digit rate will suffice, we seek to find the smallest number of messages whose total probability is at least $1 - \delta$. These would include all of the messages with probabilities greater than those of the typical set, (8-29), whose total probability can be no more than $\delta$, plus a sufficient number of typical messages to make up the remainder of $1 - \delta$. Since the probabilities of the latter do not exceed $b^{-[H'(\mathbf{X})-\epsilon]T}$, we must include at least $(1 - 2\delta)b^{[H'(\mathbf{X})-\epsilon]T}$ messages.

Taking the logarithm to the base $b$, we see that the code-word length $L$ must be at least $[H'(\mathbf{X}) - \epsilon]T + \log (1 - 2\delta)$ to provide enough different code words to permit correct decoding with a probability of at least $1 - \delta$. Dividing by $T$ and passing to the limit $T \to \infty$, we see that no code-digit rate $L/T$ less than the entropy rate $H'(\mathbf{X})$ will suffice for reliably decodable coding of the information-source output. This is the second half of the fundamental coding theorem for a discrete information source.

**Variable-Length Coding.** The coding just described was not able to provide distinct code words for the messages whose probabilities are less than those of the typical set, (8-28). This shortcoming is not serious, since for any fixed $\epsilon > 0$ the total probability of such messages vanishes as $T \to \infty$. However, it can be remedied by using longer codes for such messages, as in the Shannon-Fano coding[1] to be described next. Because of its rare use, the added length does not increase the average code-digit rate. However, it results in an uneven flow of code digits when successive messages are encoded in this way, and buffer storage is needed to even it out. Whenever the store becomes empty or full, as will happen repeatedly, the even flow of code digits can no longer be maintained, and time will be wasted or information lost.

Such variable-length coding can be done by listing all possible information-source outputs or "messages" of duration $T$ in order of

[1] C. E. Shannon, *op. cit.*, sec. 9; R. M. Fano, The Transmission of Information, I, *MIT Res. Lab. Electron. Tech. Rept.* 65 (1949).

nonincreasing probability, from the most likely to the least likely. Denoting by $p_i$ the probability of the $i$th message in the list, we define $P_1 = 0$ and $P_{i+1} = P_i + p_i$, so that $P_i$ is the cumulative probability of all messages up to the $i$th, and we express the number $P_i$, which lies between 0 and 1, in the usual $b$-ary (e.g., binary or decimal) representation. The code word for the $i$th message, then, will be the first $L_i = -[\log p_i]$ digits of this representation; these brackets denote the largest integer that does not exceed the quantity within them. Thus $L_i$ is a nondecreasing function of $i$, the more likely messages being represented by shorter code words and the less likely by longer ones.

Since $p_i \geq b^{-L_i}$, we see that $P_{i+1}, P_{i+2}, \ldots$ all differ from $P_i$ by at least $b^{-L_i}$, and hence the code words for the messages beyond the $i$th not only are at least as long as that for the $i$th message, but cannot begin with the same $L_i$ digits. Thus, receiving one-by-one the digits of the code word for the $j$th message, we would never find that they constituted the complete code word for any message until we received all $L_j$ of them, and we would then immediately recognize that we had come to the end of a code word, since no other code word starts with the same $L_j$ digits. As a result, if the code words for different messages are written one after another without any commas to separate them, they can nevertheless be separated unambiguously. Codes which, like the present and previous ones, have this property, are called *comma-free.*

Since $-\log p_i \leq -[\log p_i] < 1 - \log p_i$, the average code-word length $L = \Sigma_i p_i L_i$ lies between the message entropy $H(\mathbf{X}) = -\Sigma_i p_i \log p_i$ and $H(\mathbf{X}) + 1$. Since both of these, when divided by $T$, become $H'(\mathbf{X})$ in the limit $T \to \infty$, we see that the code-digit rate $H'(\mathbf{X})$ suffices to encode the atypical as well as the typical messages.

By a different technique,[1] the average code-word length $L$ can be reduced to its minimum possible value, but the code-digit rate cannot be reduced below the entropy rate $H'(\mathbf{X})$ of the source if there is to be a one-to-one correspondence between the messages and the code sequences, since both must have the same entropy rate. Codes like these, which use no more digits per second than necessary may be called *minimum-redundancy* or *irredundant.* The notion of "redundancy" will shortly be made more precise.

**Coding and Gambling.** The fundamental coding theorem for a discrete information source provides a concrete interpretation of entropy rate, as contrasted with its appeal as a measure of uncertainty. John L.

[1] D. A. Huffman, A Method for the Construction of Minimum-Redundancy Codes, *Proc. IRE,* **40**:1098–1101 (1952) and Willis Jackson (ed.), "Communication Theory," pp. 102–110, Academic Press Inc., New York, and Butterworth & Co. (Publishers), Ltd, London, 1953.

Kelly, Jr.,[1] has shown it to have yet another concrete interpretation, unrelated to coding, which, however, has not yet had much consequence. He has shown that the information rate $I'(\mathbf{X};\mathbf{Y})$ is the expected rate of increase of the logarithm of a gambler's capital when he bets on the $x$ sequence at a priori fair odds with the help of the $y$ information received over a private wire, betting in such a way as to maximize this expected rate.

**Problem 8-16.** (*a*) What would be the entropy rate and utility of the unitary ($b = 1$) code in which the $i$th most likely message is represented by $i$ zeros? (*b*) Determine the Shannon-Fano code for a sequence of one, two, three, and four tosses of a biased coin whose probability of heads is $\frac{2}{3}$. Compare its average word length with the entropy, noticing that the coding efficiency (their ratio) improves as the message length grows.

## REDUNDANCY

If the output of a discrete, ergodic information source is a sequence of selections from an alphabet (including, for example, letters, punctuation, and space) of $A$ symbols, there are $A^N$ possible sequences of (block) length $N$. If all of these were equally probable, the entropy of the sequence would be, by (8-7), $N \log A$, and the entropy rate of the source would be $\log A$ per symbol.

In general, however, not all sequences are equally likely, and, by (8-24) and (8-30), the number of possible sequences may be regarded as roughly $b^{H(\mathbf{X})}$ for large $N$, where $H(\mathbf{X}) = H(X_1,X_2, \ldots ,X_N)$ is the entropy of the sequence of $N$ symbols. In the case of natural languages, for example, not only are some letters used more frequently than others, but there are also statistical dependences between successive letters. These are strongest in the case of neighboring letters, as with $q$, which is almost invariably followed by $u$ (the $u$ therefore conveying no information), and $j$, which is almost never followed by a consonant in English, but the statistical dependence extends over several successive letters and, when meaning is taken into account, over much longer spans. We suppose, however, that the statistical influence disappears as the span lengthens further and that the sequence of symbols can be regarded as ergodic. By means of ergodic processes, C. E. Shannon[2] and N. M. Abramson[3] have generated sequences resembling English messages,

[1] J. L. Kelly, Jr., A New Interpretation of Information Rate, *IRE Trans. Inform. Theory,* **IT-2**:185–189 (1956); and *Bell System Tech. J.*, **35**:917–926 (1956).

[2] C. E. Shannon, *op. cit.*, sec. 3, pp. 388–389.

[3] N. M. Abramson, "Information Theory and Coding," sec. 2-8, McGraw-Hill Book Company, New York, 1963. The importance of this work is not that it estab-

thereby lending plausibility to this hypothesis. Only to the extent that the sequence is unpredictable, of course, is it necessary to communicate it. (We ignore the fact that the information it conveys may, like much advertising or foreign language, turn out to be of no interest or meaning for the ultimate recipient, as such considerations are beyond the scope of communication theory.)

The difference between $H(\mathbf{X})$ and its maximum possible value, $N \log A$, is called the *redundancy* of the sequence of $N$ symbols, $D(\mathbf{X})$, and $D'(\mathbf{X}) = \log A - H'(\mathbf{X})$ is the redundancy rate (per symbol).[1] The redundancy expresses how much the constraints of the language and of the meaning restrict the sequences. The quotient $D(\mathbf{X})/(N \log A)$ or $D'(\mathbf{X})/\log A$ is called the *relative redundancy* and is expressed in percent. Since, as we have seen, a long sequence of $N$ symbols can be represented by about $H(\mathbf{X})$ $b$-ary code digits and hence by about $H(\mathbf{X})/\log A$ symbols selected from an alphabet of size $A$ (since $\log_A P = \log_b P/\log_b A$), the difference between $N$ and $H(\mathbf{X})/\log A$ represents the number of unnecessary (redundant) symbols.

**The Redundancy of English.** By the use of several different approaches, Shannon[2] has shown that, for ordinary English messages, when constraints acting over a span of about eight letters are taken into account, the relative redundancy is about 50 percent. When longer-range interactions and meaning are taken into account, the redundancy is found to rise to about 75 percent, but this figure varies with literary style and has been known to approach 100 percent on occasion. Some writers who reject the usual constraints on vocabulary, sentence structure, and the like, achieve a redundancy below 75 percent and are thus able, perhaps, to convey an unusually large number of ideas per word.

The 50-percent figure can be obtained by applying (8-8) to the observed statistics of the possible combinations of letters. Another approach used by Shannon simplifies the computation. A person, armed with tables of the statistics of letter combinations and words, is asked to guess the successive letters of a preselected message. If his $r$th guess of a letter is correct, he is asked to go on to the next letter of the message. The sequence of numbers $r$, taken together with the guessing scheme, determines the message uniquely and so must have the same entropy. The

---

lishes the (questionable) ergodicity of English messages but that it suggests that a communication system capable of satisfactorily handling the output of an appropriate ergodic information source should suffice for the transmission of English messages.

[1] We are in effect taking the average symbol duration $1/\nu$ as our unit of time here, and so $\nu = 1$.

[2] C. E. Shannon, Prediction and Entropy of Printed English, *Bell System Tech. J.*, **30**:50–64 (1951).

entropy can then be found by applying (8-8) to the observed statistics of the numbers $r$, which tend to be more independent of each other than are the letters of the original message.

Another indication of a 50-percent redundancy is the fact that it is usually just barely possible to reconstruct an English message from which half the symbols have been deleted. The fact that two-dimensional crossword puzzles can be constructed only with difficulty also supports the 50-percent figure, since only a fraction roughly $b^{H(\mathbf{X})}/A^N$ of the $A^N$ possible ways of filling in a diagram having $N$ spaces will yield words in the horizontal direction, a similar fraction holding in the vertical direction. If these two effects were independent, only a fraction $(b^{H(\mathbf{X})}/A^N)^2$ of the $A^N$ possible filled-in diagrams would be acceptable, i.e., roughly $b^{2H(\mathbf{X})}/A^N$ or, by (8-24), $b^{[2H'(\mathbf{X})-\log A]N}$ altogether. If this number is not to be extremely small or extremely large for large $N$, $H'(\mathbf{X})$ must be $\frac{1}{2} \log A$, and the redundancy must be 50 percent.

This figure is also corroborated by the observation that, when a running key (i.e., a random, unending English text) is used to encipher an unknown English message—for example, by "adding" corresponding letters of the key and the message to get the corresponding letter of the cryptogram, with $a + b = c$, $a + c = d$, and so on—it is sometimes possible to deduce both the message and the key from the cryptogram. In this case, the cryptogram, whose entropy rate may be nearly $\log A$ per symbol, conveys information at a rate equal to twice the entropy per symbol of English, and this entropy rate $H'(\mathbf{X})$ is therefore again about $\frac{1}{2} \log A$. If it were greater, such decryption would not be possible; if it were less, it would be relatively easy.

**The Usefulness of Redundancy.** The redundancy of natural languages is often helpful in correcting errors that arise in the transmission of information—notably in handwritten messages. It also serves to identify the language of the message so that it will be interpreted correctly, thereby establishing that the ideas conveyed come from the message source rather than from the method of interpretation. This notion will be elucidated in connection with cryptanalysis, but it is also relevant to the interpretation of poetry.

However, redundancy is not *necessarily* helpful in correcting errors. If a space should be inserted after every symbol of a message, the resulting sequence would have a 50-percent redundancy in consequence of these spaces (in addition to that of the original message), but the spaces could not help to correct errors that affect successive symbols independently. The redundancy would be the same if, instead, each message symbol were repeated, but the repetitions would generally help to correct errors. Thus, if redundancy is to be useful, it must be suited to the situation in which it is to be used.

Where redundancy is not needed, it can be removed and the information rate thereby increased by coding the message in one of the ways previously described, for example. The resulting representation becomes irredundant as the coding block length becomes infinite.

**Equivocation and Prevarication.** If, as a result of some random disturbance, the sequence $\mathbf{y}$ is received instead of the intended message $\mathbf{x}$, we may call the consequent uncertainty as to the message $H(\mathbf{X}|\mathbf{y})$ or $H(\mathbf{X}|\mathbf{Y})$ the *equivocation*. Likewise, the uncertainty $H(\mathbf{Y}|\mathbf{x})$ or $H(\mathbf{Y}|\mathbf{X})$ as to what will be received when a given message is sent may be called the *prevarication* (other names have also been used for this conditional entropy).

From (8-14) we see that the average equivocation can be expressed in terms of the average prevarication as

$$H(\mathbf{X}|\mathbf{Y}) = H(\mathbf{Y}|\mathbf{X}) + H(\mathbf{X}) - H(\mathbf{Y}).$$

If $\mathbf{x}$ and $\mathbf{y}$ are both sequences of $N$ symbols selected from an alphabet of size $A$, then $H(\mathbf{Y}) \leq N \log A$, and hence

$$H(\mathbf{X}|\mathbf{Y}) \geq H(\mathbf{Y}|\mathbf{X}) + H(\mathbf{X}) - N \log A = H(\mathbf{Y}|\mathbf{X}) - D(\mathbf{X}), \quad \text{(8-32)}$$

where $D(\mathbf{X}) = N \log A - H(\mathbf{X})$ is the message redundancy. Thus the equivocation cannot be less than the difference between the prevarication and the redundancy.[1]

In many communication systems the prevarication, which is due to errors in transmission, is independent of the transmitted message. The redundancy, on the other hand, depends only on the statistics of the language or code of the message. From (8-32) we see that unequivocal (i.e., substantially error-free) communication requires that the redundancy of the transmitted message be no smaller than the prevarication of the channel. Fulfillment of this condition, however, does not guarantee reliable communication unless, as we shall see later, the redundancy takes a suitable form.

The importance of redundancy in combating errors has long been known. Since there is almost no redundancy in the numbers appearing in telegrams, any error probability, however small, makes the numerical parts of messages ambiguous. Numbers, therefore, are either spelled out or are repeated, thereby introducing a redundancy comparable to that of the rest of the text.

**Problem 8-17.** The symmetric binary channel (Problem 8-10) is used $N$ times to transmit a message of $N$ binary symbols, with a statistically independent result for each use. Show that, if the error probability $1 - P$ lies

[1] N. M. Blachman, Prevarication vs. Redundancy, *Proc. IRE,* **50**:1711–1712 (1962).

between 0.1107 and 0.8893, reliable transmission of messages having a 50-percent redundancy is impossible. Show that, if the message symbols are taken in groups of five to represent the 32 teletype characters, the latter are received with error probability 0.441 when $P = 0.8893$. Although about four characters out of nine will then be wrong, they will convey information about the transmitted message because the correct character is likely to be one of the five differing in only one binary symbol from the incorrect one.

## CRYPTOGRAPHY

When a secret message[1] $\mathbf{x}$ is to be sent by some means such as radio which permits interception by an interloper (Fig. 8-3), it is first transformed according to a set of rules known to both sender and receiver into

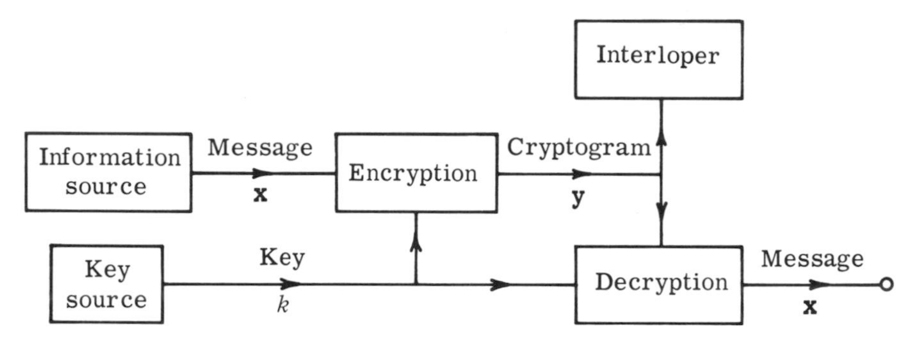


*Figure 8-3.* Block diagram of a secrecy system.

a cryptogram $\mathbf{y}$. We shall suppose that $\mathbf{x}$ and $\mathbf{y}$ are jointly ergodic sequences of $N$ symbols from an alphabet of size $A$. The message and cryptogram alphabets can be different, but we shall assume them to have equal numbers of symbols. The cryptographic transformation, for example, might be a simple substitution cipher, with each symbol in the message alphabet represented by a fixed symbol in the cryptogram alphabet. We shall restrict our attention to ciphers, which are able to transform *any* sequence $\mathbf{x}$, and shall ignore codes, which transform only words, phrases, or whole messages.

The number of possible simple substitution ciphers is $A!$, for the first letter of the message alphabet can be represented by any of the $A$ letters of the cryptogram alphabet, the second by any of the remaining $A - 1$, and so on. With $A = 26$ this number is about $4 \times 10^{26}$. The cipher is thus specified by any one of the $A!$ permutations of the alphabet, which serves as the "key."

[1] C. E. Shannon, Communication Theory of Secrecy Systems, *Bell System Tech. J.*, **28**:656–715 (1949).

We suppose that the interloper knows the kind of cipher being used, e.g., simple substitution, but not the key. However, we assume that he knows the prior probabilities of the possible keys as well as the statistics of the language, which determine the prior probabilities of the possible messages. From the intercepted cryptogram $\mathbf{y}$, along with this information, he tries to deduce the key $k$ or the message $\mathbf{x}$.

**Key and Message Equivocation.** Because any key establishes a one-to-one correspondence between messages and cryptograms, we have $H(\mathbf{Y}|K) = H(\mathbf{X})$. Hence, by (8-14), the equivocation of the key is

$$\begin{aligned} H(K|\mathbf{Y}) &= H(\mathbf{Y}|K) + H(K) - H(\mathbf{Y}) \\ &= H(K) - H(\mathbf{Y}) + H(\mathbf{X}) \\ &\geq H(K) - D(\mathbf{X}) \end{aligned} \tag{8-33}$$

by the same reasoning as in (8-32). For large $N$, the redundancy $D(\mathbf{X})$ is roughly $ND'(\mathbf{X})$, by (8-24), where $D'(\mathbf{X}) = \log A - H'(\mathbf{X})$ is the redundancy rate of the message (per symbol). Substituting

$$D(\mathbf{X}) = ND'(\mathbf{X})$$

into (8-33), we see that there will necessarily remain uncertainty about the key until at least roughly

$$N_1 = \frac{H(K)}{D'(\mathbf{X})} \tag{8-34}$$

symbols of the cryptogram have been intercepted. This value is called the *unicity length* of the cipher because, as we shall see, if it is a good cipher, a few more than $N_1$ symbols of the cryptogram suffice to determine the key (and hence the message) uniquely.

For a simple substitution cipher with $A = 26$, if the relative redundancy is 50 percent and all 400 septillion keys are equally likely, we have $D'(\mathbf{X}) = \frac{1}{2} \log 26$ and $N_1 = 37$ letters. A shorter cryptogram can usually be deciphered in several reasonable ways, all exhibiting the same pattern of repeated letters, but a longer one cannot. Because the simple substitution cipher makes use of only a small part of the key in enciphering each letter, the key may remain ambiguous until the rare letters of the alphabet have nearly all appeared in the message, and this generally requires a somewhat larger $N$.

To preserve the secrecy of the message for as large an $N$ as possible, i.e., to ensure a positive equivocation of the key, (8-34) should be made as large as possible. Thus the number of possible keys should be made as large as possible, and all of them should be used with equal probability in order to maximize $H(K)$. Also, the message redundancy $D'(\mathbf{X})$ should be minimized by deleting all unnecessary words and letters, for example. As we have seen, however, reduction of the redundancy makes it more difficult for the legitimate receiver to correct any errors that may have arisen in transmission.

From (8-2) we see that a joint entropy, such as $H(K,\mathbf{X})$ cannot be less than the entropy of one of its parts, for example, $H(\mathbf{X})$; hence $H(\mathbf{X}|\mathbf{Y}) \leq H(K,\mathbf{X}|\mathbf{Y})$. Since $\mathbf{y}$ and $k$ together determine $\mathbf{x}$, we have $H(K,\mathbf{X}|\mathbf{Y}) = H(K|\mathbf{Y})$ and, therefore,

$$H(\mathbf{X}|\mathbf{Y}) \leq H(K|\mathbf{Y}), \tag{8-35}$$

that is, the equivocation of the message cannot exceed the equivocation of the key.

As $N$ increases from zero, $H(K|\mathbf{Y})$ falls from its initial value $H(K)$ (Fig. 8-4), while $H(\mathbf{X}|\mathbf{Y})$ at first grows roughly as $N$, being approximately $H(\mathbf{X}) = NH'(\mathbf{X})$, until the two equivocations become comparable. Thereafter both decrease together.

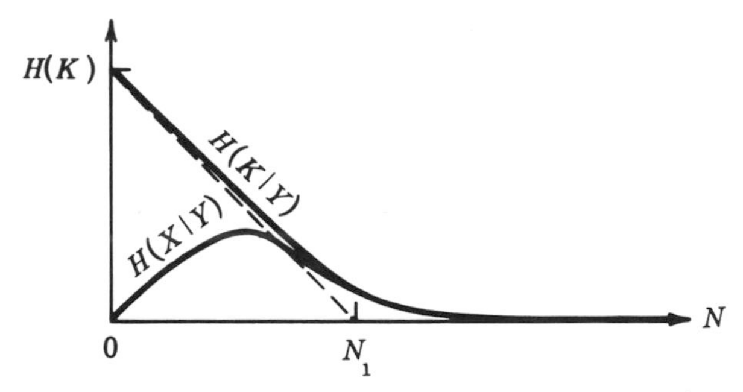


*Figure 8-4.* The equivocation of the key and of the message as functions of the length $N$ of the intercepted cryptogram. The dashed line is the lower bound (8-33) for the equivocation of the key.

**Exhaustive Cryptanalysis.** To wring all available information about the key and the message out of the intercepted cryptogram, an interloper might, in principle, try every possible key in a decryption device like that of the legitimate receiver, their number being $b^{H(K)}$ if all have equal prior probability. (It would take him ten billion years to run through 26! keys at the rate of a billion per second; there are easier ways to break a simple substitution cipher.) One of the resulting decipherments will be the correct message, and the others, if the cipher is a good one, will appear to be independent random sequences of $N$ symbols, giving no hint as to the actual message. (In this respect, the simple substitution cipher is not a very good one.)

If the number of keys, $b^{H(K)}$, exceeds $A^N$, the number of possible sequences of length $N$, the interloper will have learned nothing about the message, for he will have got all possible messages of length $N$, or at least most of them. In other words, for $N < H(K)/\log A$ the equivocation $H(\mathbf{X}|\mathbf{Y})$ of the message equals $H(\mathbf{X})$, which is approximately $NH'(\mathbf{X})$. Nevertheless, he will have learned something about the *key*, for he can discard all those keys that produced meaningless sequences of letters.

With a probability approaching unity, the true message will, as a result of the language constraints, be one of roughly $b^{H(\mathbf{X})}$ different sequences of length $N$. These represent a fraction $b^{H(\mathbf{X})}/A^N = b^{-D(\mathbf{X})}$ or approximately $b^{-ND'(\mathbf{X})}$ of the $A^N$ such sequences. Hence the expected number of keys in addition to the actual one that will remain is $b^{H(K)-D(\mathbf{X})}$. By (8-28) the logarithms of the posterior probabilities of all of these keys will be approximately equal, and hence, if this number is large, i.e., if $N < N_1$, the equivocation of the key is $H(K) - D(\mathbf{X})$ or

$$H(K|\mathbf{Y}) = H(K) - ND'(\mathbf{X}). \tag{8-36}$$

When the number of keys $b^{H(K)}$ is small compared to the number of different sequences $A^N$ of length $N$, that is, when $N > H(K)/\log A$, each plausible message probably results from only a single key, and hence in this range the message and the key have the same equivocation,

$$H(\mathbf{X}|\mathbf{Y}) = H(K|\mathbf{Y}).$$

When $N$ exceeds the unicity length, $b^{H(K)-ND'(\mathbf{X})}$ is less than unity and is approximately the probability that there is a second key (besides the correct one) that yields a reasonable message. (The probability that there are more than two may be neglected.) This is therefore the probability that there is a one-bit posterior equivocation of the key, its equivocation being zero otherwise. The average of these two posterior equivocations is

$$H(K|\mathbf{Y}) = b^{H(K)-ND'(\mathbf{X})} \tag{8-37}$$

bits, which is also $H(\mathbf{X}|\mathbf{Y})$. Thus, beyond the unicity length, the equivocation of the key and message falls exponentially, being multiplied by 0.2 for each increase of $N$ by one letter when $A = 26$ and the message redundancy is 50 percent.

This rapid decrease in equivocation beyond the unicity length means that very few more than $N_1$ symbols of the cryptogram are needed to make possible the determination of the message, at least in principle, but if the cipher is good, it will require a great deal of work. Such security is obtained at the cost of effort required in enciphering and deciphering by those who know the key, and at the risk of thorough garbling if any symbol of the cryptogram is received incorrectly.

*Caveat.* When fewer than $N_1$ symbols of the cryptogram are intercepted, there are many keys that will yield different plausible messages. In this case it is not sufficient to find a key that gives a reasonable message, and any solution obtained with $N < N_1$ is necessarily suspect. Care must be exercised, too, in evaluating the $H(K)$ in (8-34), for if the nature of the cipher is not known in advance, $H(K)$ must include the prior uncertainty as to the cipher as well as that of the key.

**Problem 8-18.** Find $N_1$ for a digram substitution cipher with a 26-letter alphabet and 50-percent redundancy. Here the message is divided into pairs of letters, and each pair is replaced by another pair of letters according to a fixed scheme, as in simple substitution.

**Problem 8-19.** Determine the unicity length for a transposition cipher of period 5 with 26-letter alphabet and 50-percent redundancy. Here the message is divided into sets of five successive letters, and each set is permuted within itself in the same way. The absurdly small answer results from the fact that, with a transposition cipher, the letters retain their individual probabilities, and these are of no use, therefore, in eliminating keys.

## □ COMMUNICATION THROUGH A NOISY CHANNEL

A noisy communication channel changes its input sequence $\mathbf{x}$ into another, output sequence $\mathbf{y}$ in a random way, similar to that in which encryption changes a message, so far as the interloper is concerned. Such a channel is described by the matrices of conditional probabilities $P(\mathbf{y}|\mathbf{x})$ that any particular output sequence $\mathbf{y}$ of arbitrary duration $T$ will result when the input sequence during the same interval is any given $\mathbf{x}$. The description of a communication channel must also specify what sequences are acceptable as inputs; in other words, it must indicate the input language, power level, peak voltage, or the like. We shall confine our attention to ergodic channels, i.e., to channels whose input and output are jointly ergodic whenever the input is ergodic.

**Overcoming the Noise.** Given any ergodic input statistics $\mathbf{P}(\mathbf{x})$, we can, by using the channel statistics $\mathbf{P}(\mathbf{y}|\mathbf{x})$, determine the output statistics $\mathbf{P}(\mathbf{y})$ and the (posterior) distribution $\mathbf{P}(\mathbf{x}|\mathbf{y})$ of the input $\mathbf{x}$ for any given output sequence $\mathbf{y}$, as well as the input entropy $H(\mathbf{X})$, the equivocation $H(\mathbf{X}|\mathbf{Y})$, and the mutual information

$$I(\mathbf{X};\mathbf{Y}) = H(\mathbf{X}) - H(\mathbf{X}|\mathbf{Y})$$

for sequences of duration $T$. For large $T$, as (8-30) indicates, the input sequence is very likely to be one of roughly $b^{H(\mathbf{X})}$ different sequences, and an output sequence $\mathbf{y}$ is very likely to have resulted from one of roughly $b^{H(\mathbf{X}|\mathbf{y})}$ different input sequences. If $\mathbf{y}$ is a random output $H(\mathbf{X}|\mathbf{y})$ is very likely, because of the law of large numbers, to be relatively close to its average, $H(\mathbf{X}|\mathbf{Y})$, and the latter number becomes roughly $b^{H(\mathbf{X}|\mathbf{Y})}$.

Likewise, the output of an information source (Fig. 8-5) of entropy rate $R$ during the time $T$ is very likely to be one of roughly $b^{RT}$ different messages, i.e., within a factor whose logarithm is of a smaller order than $T$. Just as it is possible for an interloper to decipher correctly a message encrypted by means of an unknown key, it is possible to reconstruct a message that has been garbled by transmission through a noisy channel, provided that a sufficient amount of redundancy of the right sort has been incorporated. To do this the message $m$ is suitably encoded by the transmitter as a signal $\mathbf{x} = \mathbf{f}(m)$, which serves as the channel input, and the receiver performs the decoding of the received signal $\mathbf{y}$ to recover the message $m = g(\mathbf{y})$.

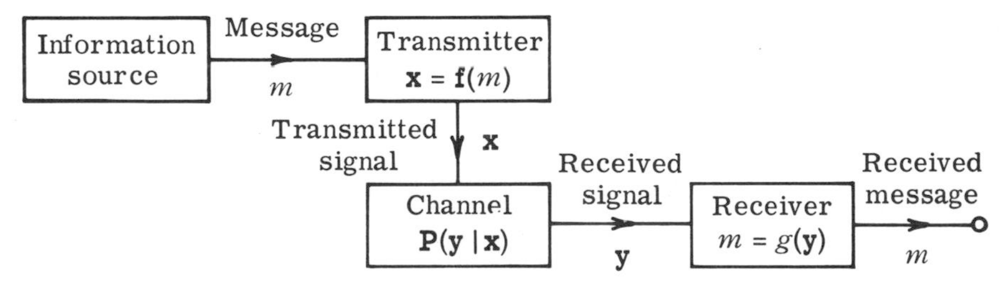


*Figure 8-5.* Block diagram of a communication system. The information-source output—the message $m$—is encoded by the transmitter as the signal $\mathbf{x} = \mathbf{f}(m)$, which serves as the input sequence for the channel with probability matrix $\mathbf{P}(\mathbf{y}|\mathbf{x})$. The channel's output sequence $\mathbf{y}$ is decoded by the receiver, whose output is hopefully the original message $m = g(\mathbf{y})$.

If the penalties for all errors are equal, the receiver should always decide in favor of the message with the highest posterior probability, but it is much more difficult to say what sort of coding it would be best for the *transmitter* to use. Nevertheless, it is easy to show that it is *possible* for the transmitter to incorporate the necessary redundancy, provided only that the information rate $R$ of the source is less than an information rate $I'(\mathbf{X};\mathbf{Y})$ that can be attained by the channel for some ergodic input statistics $\mathbf{P}(\mathbf{x})$.

To do this we denote by $H(\mathbf{X})$ the entropy of any such distribution $\mathbf{P}(\mathbf{x})$, and we notice that for large $T$ the roughly $b^{H(\mathbf{X})}$ sequences $\mathbf{x}$ of the associated typical set account for almost the entire distribution. We denote by $H(\mathbf{X}|\mathbf{Y}) = H(\mathbf{X}) - I(\mathbf{X};\mathbf{Y})$ the equivocation that would result if the channel input had the statistics $\mathbf{P}(\mathbf{x})$. With each output sequence $\mathbf{y}$ we can associate the typical set of input sequences $\mathbf{x}$ that could reasonably have produced it if the input distribution was $\mathbf{P}(\mathbf{x})$; with a probability approaching unity the number of input sequences in this set will then be roughly $b^{H(\mathbf{X}|\mathbf{Y})}$.

We proceed by choosing a code word at random from among the $b^{H(\mathbf{X})}$ input sequences to represent each of the $b^{RT}$ messages. When a message $m$ is encoded in this way as a signal $\mathbf{x}$ and the resulting channel output is $\mathbf{y}$, $\mathbf{x}$ is very likely to be among the $b^{H(\mathbf{X}|\mathbf{Y})}$ input sequences which could reasonably have produced $\mathbf{y}$, and the receiver, knowing the code used by the transmitter, has no difficulty in deciding that the message $m$ was sent, as long as there is no other message whose code word falls among the same set of $b^{H(\mathbf{X}|\mathbf{Y})}$ input sequences associated with $\mathbf{y}$. Since the probability that the code word for any message falls within this set is $b^{H(\mathbf{X}|\mathbf{Y})}/b^{H(\mathbf{X})} = b^{-I(\mathbf{X};\mathbf{Y})}$, the probability that it does not is $1 - b^{-I(\mathbf{X};\mathbf{Y})}$, which, for large $T$ and, consequently, large $I(\mathbf{X};\mathbf{Y})$, is equal to $\exp_e(-b^{-I(\mathbf{X};\mathbf{Y})})$, being the first two terms of its power-series expansion. Hence the probability that there is *no* other message whose code word falls within the same set of input sequences is

$$[\exp_e(-b^{-I(\mathbf{X};\mathbf{Y})})]^{b^{RT}-1} = \exp_e(-b^{RT-I(\mathbf{X};\mathbf{Y})} + b^{-I(\mathbf{X};\mathbf{Y})}).$$

As $T$ grows infinite this probability approaches unity, provided that the information rate $R$ of the message source is less than the rate of growth of $I(\mathbf{X};\mathbf{Y})$, that is, the information rate of the channel when its input has the ergodic statistics $\mathbf{P}(\mathbf{x})$ that we have so far regarded as given. Notice that by $\mathbf{P}(\mathbf{x})$ we do *not* mean the input statistics of the channel when it is fed by the transmitter of Fig. 8-5.

Hence, as long as

$$R < I'(\mathbf{X};\mathbf{Y}), \tag{8-38}$$

the mean error probability, averaged over all possible ways of encoding, approaches zero as $T$ grows infinite. Since the error probability can never be negative, it follows that there exist codes with arbitrarily small probabilities of incorrect reception.[1] Moreover, it also follows that the overwhelming proportion of all ways of assigning signals to messages for large $T$ yield low error probabilities, for otherwise the mean error probability could not be small. However, there is a small probability that a randomly chosen code will not yield a small error probability, and so it is not appropriate to use such a code—or a code for *any* kind—without first verifying that that particular code yields a satisfactorily low error probability.

[1] We seem not only to have proved this but also to have obtained an approximate upper bound on the error probability for the best code using signals of duration $T$. However, our approximations make this bound extremely rough; for precise bounds, see, for example, C. E. Shannon, Certain Results in Coding Theory for Noisy Channels, *Information and Control*, **1**:6–25 (1957), and R. G. Gallager, A Simple Derivation of the Coding Theorem and Some Applications, *IEEE Trans. Inform. Theory*, **IT-11**: 3–18 (1965). To obtain such bounds, additional assumptions are required as to the nature of the channel.

**Problem 8-20.** Show that there exist codes having not only arbitrarily small error probabilities but also arbitrarily low rates of equivocation. Do this by using the same signal to represent all atypical messages and noting that, when the message is received incorrectly, its posterior entropy does not exceed its prior entropy $RT$ plus the logarithm of the number of typical messages, also $RT$.

**The Fundamental Theorem on the Noisy Channel.** For given channel statistics $\mathbf{P}(\mathbf{y}|\mathbf{x})$ and restrictions on the input $\mathbf{x}$, the greatest value that $I'(\mathbf{X};\mathbf{Y})$ can have, maximized over all admissible ergodic input statistics $\mathbf{P}(\mathbf{x})$, is called the *channel capacity*

$$C = \sup_{\mathbf{P}(\mathbf{x})} I'(\mathbf{X};\mathbf{Y}), \tag{8-39}$$

the supremum being the least upper bound on the mutual information rate for all coding block lengths $T$ and all $\mathbf{P}(\mathbf{x})$. Using this best possible value for the right-hand side of (8-38), we obtain the first half of the *fundamental theorem on the noisy channel*: By means of suitable coding it is possible to transmit information from a source of rate $R$ through a channel of capacity $C$ with an arbitrarily small error probability, provided that $R < C$.

The second half of the theorem states that this is not possible when $R > C$. We shall prove it by first noticing that, by (8-18) and (8-39), with $m$ the message coded as $\mathbf{x}$,

$$\begin{aligned} I'(M;\mathbf{Y}) &= H'(M) - H'(M|\mathbf{Y}) \\ &= H'(\mathbf{Y}) - H'(\mathbf{Y}|M) \\ &\leq H'(\mathbf{Y}) - H'(\mathbf{Y}|M,\mathbf{X}) \\ &= H'(\mathbf{Y}) - H'(\mathbf{Y}|\mathbf{X}) \\ &\leq C, \end{aligned} \tag{8-40}$$

since $P(\mathbf{y}|m,\mathbf{x}) = P(\mathbf{y}|\mathbf{x})$. In other words, the mutual information rate of two statistically independent channels in tandem, such as the transmitter and the communication channel of Fig. 8-5, cannot exceed the capacity of either channel. Since $H'(M) = R$, it follows that

$$H'(M|\mathbf{Y}) \geq R - C, \tag{8-41}$$

that is, the equivocation rate of the message is no less than the amount by which the information rate of the message source exceeds the channel capacity.

If we eliminated from consideration all atypical messages, the posterior entropy of the message could not exceed $RT$ in those cases in which the receiver output is the wrong message. Hence, breaking down the choice of receiver output message for any received signal $\mathbf{y}$ into a choice between the right (intended) message and a wrong one followed, in the latter case, by the choice of a particular wrong message, we see that the equivocation of the message equals the sum of the entropy of the binary choice plus the error probability times something not exceeding $RT$, by (8-2). Dividing by $T$ and averaging over all $m$ and $\mathbf{y}$, we see that, if the error probability could be made arbitrarily small, so could the equivocation rate, contrary to (8-41). Thus the error probability cannot be made arbitrarily small when $R > C$, for the inclusion of the atypical messages could only make reliable transmission more difficult.

The foregoing proof of the fundamental theorem on the noisy channel follows the lines of Shannon's original proof.[1] Later he published a more rigorous though more specialized proof.[2] Feinstein was the first to give a proof[3] not depending on intuition; he used McMillan's rigorous proof[4] of the "asymptotic equipartition property" of ergodic information sources (8-28), by which their output sequences can be separated into the typical and atypical sets.

Wolfowitz[5] has not only proved the theorem for a variety of channels but has also strengthened its second half in a number of cases, obtaining a "strong converse," which states that, for $R > C$, the error probability not only cannot be made arbitrarily small but, moreover, necessarily approaches unity as the coding block length grows infinite.

**Channel Capacity.** Although the channel capacity (8-39) is easy to define, it is in general not at all easy to compute. Even for a discrete, memoryless channel—one whose output symbol at any time depends statistically only on the input symbol at the same time and not on any preceding input or output symbols—the calculation is far from straightforward (see Prob. 8-24), but when the input alphabet is small or the channel is symmetric, as in the case of the symmetric binary channel (Prob. 8-10), it is easy to find $C$ and the input statistics $\mathbf{P}(\mathbf{x})$ that attain it.

In general, the unaveraged prevarication $H(\mathbf{Y}|\mathbf{x})$ depends on $\mathbf{x}$ and on the channel statistics $\mathbf{P}(\mathbf{y}|\mathbf{x})$, but not on the input statistics $\mathbf{P}(\mathbf{x})$. In the case of a symmetric channel, it is the same for all $\mathbf{x}$, and its average

[1] C. E. Shannon, A Mathematical Theory of Communication, *Bell System Tech. J.*, **27**:379–423, sec. 13 (1948).

[2] C. E. Shannon, Certain Results in Coding Theory for Noisy Channels, *Information and Control*, **1**:6–25 (1957).

[3] Amiel Feinstein, "Foundations of Information Theory," McGraw-Hill Book Company, New York, 1958. See also R. M. Fano, "Transmission of Information," M.I.T. Press and John Wiley & Sons, Inc., New York, 1961.

[4] Brockway McMillan, "The Basic Theorems of Information Theory," *Ann. Math. Statist.*, **24**:196–219 (1953).

[5] Jacob Wolfowitz, "Coding Theorems of Information Theory," Springer-Verlag, Berlin, and Prentice-Hall, Inc., Englewood Cliffs, N.J., 1961; 2d ed., 1964.

value $H(\mathbf{Y}|\mathbf{X})$, therefore, is independent of $\mathbf{P}(\mathbf{x})$. Hence

$$I(\mathbf{X};\mathbf{Y}) = H(\mathbf{Y}) - H(\mathbf{Y}|\mathbf{X})$$

is maximized by that $\mathbf{P}(\mathbf{x})$ which maximizes $H(\mathbf{Y})$.

In the case of the memoryless symmetric binary channel, $H(\mathbf{Y})$ is maximized by ensuring that all $2^N$ of the possible output sequences of length $N$ have equal probability, and this can evidently be done by making each input symbol an independent random choice between the two alternatives, each with probability $\frac{1}{2}$. As a result, the output entropy rate is $H'(\mathbf{Y}) = 1$ bit/symbol. Subtracting the prevarication rate

$$H'(\mathbf{Y}|\mathbf{X}) = -P \log P - (1 - P) \log (1 - P),$$

where $1 - P$ is the error probability for each symbol sent through the channel (see Prob. 8-15), we thus get the channel capacity

$$C = P \log \frac{b}{P} + (1 - P) \log \frac{b}{1 - P},$$

where $b$ is again the base of our logarithms. Notice that this capacity is one bit per symbol for $P = 1$ or $0$ and is zero for $P = \frac{1}{2}$.

**Problem 8-21.** Find the capacity $C$ of the unsymmetric memoryless binary channel that conveys each 1 correctly with probability $P$ and each 0 correctly with probability $Q$. Show that, for $P = 1$ and $Q = \frac{1}{2}$, $C = \log \frac{5}{4}$ per symbol, which is attained by input probabilities $\frac{3}{5}$ for 1 and $\frac{2}{5}$ for 0.

**Problem 8-22.** What is the capacity of the memoryless symmetric binary erasure channel (Prob. 8-11)? Try to find its capacity in the unsymmetric case (see Prob. 8-21).

**Problem 8-23.** What can be said about the prevarication, the equivocation, and the capacity of a *noiseless* memoryless channel, i.e., one for which every $P(y_j|x_i)$ is either 1 or 0? How is the capacity attained? (Here $x_i$ and $y_j$ denote symbols of the input and output alphabets, respectively.)

**Problem 8-24.** Show that, if the matrix of probabilities $P(y_j|x_i)$ of a discrete memoryless channel is nonsingular, the output entropy rate per symbol $H'(\mathbf{Y})$ is a concave function[1] of the output probabilities $P(y_j)$, whose sum must be 1, and show that the prevarication rate $H'(\mathbf{Y}|\mathbf{X})$ is a linear function of the output probabilities. Show that the channel capacity is therefore realized by obtaining the set of $P(y_j)$ at which the plane tangent to the $H'(\mathbf{Y})$ surface is parallel to the $H'(\mathbf{Y}|\mathbf{X})$ plane. If the channel matrix is singular and

[1] That is, for any two "points" or sets of $P(y_j)$, the average of the two corresponding values of $H'(\mathbf{Y})$ is no greater than the value of $H'(\mathbf{Y})$ at the middle of the line joining the two points.

has rank $r$, it can be shown[1] that all but $r$ of the input probabilities $P(x_i)$ may be set equal to zero, thereby reducing the problem to the foregoing one, since $H'(\mathbf{Y}|\mathbf{X})$ can then take any value within a convex polyhedron, but only its lowest surface can yield the channel capacity. If the output alphabet contains two symbols and the input alphabet is larger, $r$ being 2, which two input symbols should be used?

**Coding for Noisy Channels.** The problem of devising codes that achieve satisfactorily low error probabilities has received and continues to receive much attention in the literature.[2] Codes of various kinds have been studied—some of them intended to simplify the encoding at the transmitter or the decoding at the receiver,[3] and some of them with specified "error-correcting" properties, intended to combat particular kinds of noise. "Error correcting" does not mean that incorrectly received code symbols are actually corrected but that the decoding generally yields the right message despite such errors and that the errors *can* therefore be corrected if need be. It is achieved by requiring the codes for different messages to be sufficiently different that the anticipated patterns of errors will not cause confusion. This requirement limits the number of messages than can be represented by a sequence of given length, and so introduces the necessary redundancy.

A somewhat different approach, *iterative coding*,[4] is able to make the error probability arbitrarily small by means of a code which is independent of the desired error probability; the latter is determined by the allowable decoding delay at the receiver before the corrected message must be delivered to its destination.

In practice, the signal arriving at the receiver is usually continuous, like an electric voltage, rather than discrete, like an alphabetic symbol, and a sequence of decisions must be made as to the most likely of a discrete set of values or symbols in order to make the channel appear discrete. Such decisions throw away information about the posterior probabilities of the various possible symbols so that the philosophy of error-correcting codes can be applied, for it is often easier to implement

[1] C. E. Shannon, Geometrische Deutung einiger Ergebnisse bei der Berechnung der Kanalkapizität, *Nachrichtentechnische Zeitschrift,* **10**:1–4 (1957); C. E. Shannon, Some Geometric Results in Channel Capacity, *VDE Fachber.*, **19**(II):13–15 (1956) = *Nachrichtentech. Fachber*, (*N.T.F.*), **6** (1957). See also F. M. Reza, "An Introduction to Information Theory," sec. 3-17, McGraw-Hill Book Company, New York, 1961.

[2] W. W. Peterson, "Error-Correcting Codes," John Wiley & Sons, Inc., New York, 1961.

[3] J. M. Wozencraft and Barney Reiffen, "Sequential Decoding," M.I.T. Press and John Wiley & Sons, Inc., New York, 1961.

[4] P. Elias, Error-Free Coding, *IRE Trans. Inform. Theory,* **PGIT-4**:29–37 (1954); P. Elias, Coding for Noisy Channels, *IRE National Convention Record, Part* 4, 37–46 (1955).

the decoding of a discrete sequence than a decision based on the maximum posterior *message* probability. The latter approach, which does *not* discard relevant information, is sometimes called *Wagner coding*, though it is not a kind of coding at all but, rather, is maximum-posterior-probability decoding.[1]

We shall not go further into the many kinds of discrete codes and their properties, but shall instead turn now to channels whose inputs and outputs range over a continuum instead of a discrete set of values.

[1] R. A. Silverman and M. Balser, Coding for Constant-Data-Rate Systems, *IRE Trans. Inform. Theory*, **PGIT-4**:50–63 (1954); *Proc. IRE*, **42**:1428–1435 (1954); and **43**:728–733 (1955).

# chapter 9 INFORMATION IN CONTINUOUS CHANNELS

For a channel whose input $\mathbf{x}$ is discrete but whose output $\mathbf{y}$ has a continuous distribution, there is no difficulty in calculating its mutual information rate $I'(\mathbf{X};\mathbf{Y})$ in the form $H'(\mathbf{X}) - H'(\mathbf{X}|\mathbf{Y})$ by means of the formulas of Chap. 8, the equivocation rate $H'(\mathbf{X}|\mathbf{Y})$ being the average of $H(\mathbf{X}|\mathbf{y})/T$ over the distribution of $\mathbf{y}$ in the limit $T \to \infty$. However, if we try to use the expression $I(X;Y) = H(Y) - H(Y|X)$, or if $x$, too, has a continuous distribution, we find that all of our entropies are infinite because of the infinite number of different values that $y$ can take if the range of $y$ is divided into increments $dy_j$ whose widths are made to approach zero, and $I(X;Y)$ becomes indeterminate.

## DIFFERENCES AND SIMILARITIES IN THE DISCRETE AND CONTINUOUS CASES

This difficulty can be circumvented by using (8-16) to find $I(X;Y)$. There $I(X;Y)$ is seen to be the average value of the logarithm of $P(y|x)/P(y)$, for example. In the continuous case $P(y|x)$ becomes $p(y|x)\,dy$ and $P(y)$ becomes $p(y)\,dy$ in terms of the conditional and unconditional probability density functions of $y$. Being the same increment, the two $dy$'s cancel on dividing, and we have

$$\begin{aligned} I(X;Y) &= \iint p(x,y) \log \frac{p(y|x)}{p(y)}\,dx\,dy \\ &= \iint p(x,y) \log \frac{p(x|y)}{p(x)}\,dx\,dy \\ &= \iint p(x,y) \log \frac{p(x,y)}{p(x)p(y)}\,dx\,dy, \end{aligned} \tag{9-1}$$

the integrations being over the entire range of $x$ and $y$, from $-\infty$ to $\infty$.

**Redefinition of Entropy.** Since (8-8) is always infinite in the continuous case, it is desirable to define entropy differently here. We can evidently retain the equalities (8-14) expressed in Fig. 8-2 if we define $H(X)$ in the continuous case as

$$H(X) = -\int p(x) \log p(x)\, dx \tag{9-2}$$

and, analogously,

$$\begin{aligned} H(X|y) &= -\int p(x|y) \log p(x|y)\, dx, \\ H(X|Y) &= -\int\int p(x,y) \log p(x|y)\, dx\, dy, \\ H(X,Y) &= -\int\int p(x,y) \log p(x,y)\, dx\, dy, \end{aligned} \tag{9-3}$$

and so on. Notice that, although these entropies have the same additive properties [e.g., (8-2)] as in the discrete case, they cannot readily be interpreted as measures of uncertainty, for they can be negative as well as positive. As we shall see shortly, $H(Y)$ need not equal $H(X)$, even though $x$ determines $y$ uniquely and vice versa. However, such changes of variable cannot affect the value of $I(X;Y)$, which was *not* defined arbitrarily.

**Problem 9-1.** Show that, if $x$ is normally distributed with standard deviation $\sigma$, its entropy is $H(X) = \frac{1}{2} \log (2\pi e \sigma^2)$ regardless of its mean. From this result show that, if $x$ and $y$ are jointly normal with correlation coefficient $\varrho$, their mutual information is $I(X;Y) = -\frac{1}{2} \log (1 - \varrho^2)$.

**Problem 9-2.** Show that, if $a$ obeys the Rayleigh distribution with mode $\sigma$, its entropy is $H(A) = \frac{1}{2} \log (\frac{1}{2}\gamma e^2 \sigma^2)$, where $\gamma = e^C = 1.781 \ldots$ and $C = -\int_0^\infty e^{-u} \ln u\, du = 0.5772 \ldots$ is Euler's constant.

**Problem 9-3.** Show that, if $a$ is the amplitude and $\phi'$ is the rate of change of phase of a narrow-band gaussian noise at the same instant, the centroid of the noise spectrum being the reference frequency, their mutual information is $I(A;\Phi') = \frac{1}{2} \log (e^5/16\pi\gamma) = 0.365$ bit.

**Problem 9-4.** Notice that, if the distribution of $y$ is continuous, $I(x;Y)$ can*not* be expressed as $H(Y) - H(Y|x)$, since the $dy$'s do not cancel out here. To show that $H(Y) - H(Y|x)$ does not measure the amount of information that a specific $x$ gives about $y$, suppose that $x$ takes just two values, 0 and 1, with equal probability, and that $y$ is uniformly distributed between $x - \frac{1}{2}$ and $x + \frac{1}{2}$. Show that $H(Y) - H(Y|x) = 1$ bit for either $x$, and show that, if $y = z^3$, with the result that $y$ and $z$ determine each other uniquely, then $H(Z) - H(Z|x)$ is different, being $\frac{1}{2} \log \frac{4}{3}$ for $x = 0$ and $\frac{1}{2} \log 12$ for $x = 1$. Verify that nevertheless $I(X;Z) = I(X;Y)$.

**Inequalities Satisfied by the New Entropies.** The proof (8-17) that $I(X;Y) \geq 0$, with equality if and only if $x$ and $y$ are statistically independent, remains valid in the continuous case. Hence we still have the inequality (8-18) asserting that, on the average, conditioning cannot increase an entropy. As in the discrete case, it follows that

$$H(X,Y) \leq H(X) + H(Y), \tag{9-4}$$

but (8-2) no longer implies that $H(X,Y) \geq H(X)$, since $H(Y|X)$ can now be negative.

Generalizing (9-1) and (9-2) to the case of multivariate distributions, we find, analogously to (8-10), that if the $m$-component vector $\mathbf{x}$ is confined to a set $S$ of $m$-dimensional measure (length, area, volume, . . .) $V$, then

$$\begin{aligned} H(\mathbf{X}) &= -\int_S p(\mathbf{x}) \log p(\mathbf{x})\, d\mathbf{x} \\ &= \log V + \int_S p(\mathbf{x}) \log \frac{1}{Vp(\mathbf{x})}\, d\mathbf{x} \\ &\leq \log V + \int_S p(\mathbf{x}) \left[ \frac{1}{Vp(\mathbf{x})} - 1 \right] d\mathbf{x} \log e \\ &= \log V \end{aligned} \tag{9-5}$$

by (8-9), $d\mathbf{x}$ being the $m$-dimensional volume element and the integrals being $m$-fold. Since (8-9) is an equality only for $u = 1$, $H(\mathbf{X})$ can attain this maximum value only when $\mathbf{x}$ is uniformly distributed over the set $S$.

**Problem 9-5.** Show that, if the $m$ components of $\mathbf{x}$ are jointly normal with covariance matrix $\mathbf{M}$, their joint entropy is $H(\mathbf{X}) = \frac{1}{2} \log [(2\pi e)^m |\mathbf{M}|]$. From this result show that, if the $m$ components of $\mathbf{x}$ and the $n$ components of $\mathbf{y}$ are jointly normal, their mutual information is $I(\mathbf{X};\mathbf{Y}) = \frac{1}{2} \log |\mathbf{M_x}||\mathbf{M_y}|/|\mathbf{M_{xy}}|$, where $\mathbf{M_x}$, $\mathbf{M_y}$, and $\mathbf{M_{xy}}$ are the $m \times m$, $n \times n$, and $(m + n) \times (m + n)$ covariance matrices for $\mathbf{x}$, for $\mathbf{y}$, and for $(\mathbf{x},\mathbf{y})$, respectively, $\mathbf{M_x}$ and $\mathbf{M_y}$ being submatrices of $\mathbf{M_{xy}}$.

**Change of Variable.** If $\mathbf{x}$ is a single-valued function of $\mathbf{z}$ and vice versa, the ratio of the volume element $d\mathbf{z}$ to the corresponding volume element $d\mathbf{x}$ is the jacobian $J = J\left(\frac{z_1, \ldots, z_m}{x_1, \ldots, x_m}\right)$, i.e., the determinant of the matrix of derivatives $\partial z_i/\partial x_j$. Since the element of probability $p(\mathbf{x})\, d\mathbf{x}$ expressed in terms of the probability density function of $\mathbf{x}$ must equal the corresponding element $p(\mathbf{z})\, d\mathbf{z}$ expressed in terms of the (generally different) probability density function of $\mathbf{z}$, we have $p(\mathbf{z}) = p(\mathbf{x})/|J|$ and

$$\begin{aligned} H(\mathbf{Z}) &= -\int \frac{p(\mathbf{x})}{|J|} \log \frac{p(\mathbf{x})}{|J|}\, d\mathbf{z} \\ &= -\int p(\mathbf{x}) \log \frac{p(\mathbf{x})}{|J|}\, d\mathbf{x} \\ &= H(\mathbf{X}) + \int p(\mathbf{x}) \log \left| J\left(\frac{z_1, \ldots, z_m}{x_1, \ldots, x_m}\right) \right| d\mathbf{x}. \end{aligned} \tag{9-6}$$

Thus the transformation increases the entropy by the mean value of the logarithm of the jacobian. With $m = 1$, (9-6) becomes simply

$$H(Z) = H(X) + \int p(x) \log \left| \frac{dz}{dx} \right| dx. \tag{9-7}$$

Being the same for both $H(\mathbf{X})$ and $H(\mathbf{X}|\mathbf{Y})$, this change in entropy cancels out in the calculation of $I(\mathbf{X};\mathbf{Y}) = H(\mathbf{X}) - H(\mathbf{X}|\mathbf{Y})$, and we have $I(\mathbf{Z};\mathbf{Y}) = I(\mathbf{X};\mathbf{Y})$. However, it is generally not the same for $H(\mathbf{X}|\mathbf{y})$ as for $H(\mathbf{X})$, since different distributions of $\mathbf{x}$ are involved here, and hence $H(\mathbf{X}) - H(\mathbf{X}|\mathbf{y})$ is coordinate-dependent (see Prob. 9-4).

**Problem 9-6.** Use (9-7) to explain why $H(X)$ in Prob. 9-1 contains the term $\log \sigma$ and is independent of the mean. Why is $H(\mathbf{X})$ in Prob. 9-5 independent of the mean value of $\mathbf{x}$?

**Problem 9-7.** Show that, if $s = \frac{1}{2}a^2$ and $a$ is Rayleigh-distributed as in Prob. 9-2, $H(S) = \log e\sigma^2$.

**Problem 9-8.** Some authors call $J(\mathbf{x};\mathbf{Y}) = \Sigma_{\mathbf{y}} P(\mathbf{y}|\mathbf{x}) \log [P(\mathbf{y}|\mathbf{x})/P(\mathbf{y})]$ in the discrete case and $J(\mathbf{x};\mathbf{Y}) = \int p(\mathbf{y}|\mathbf{x}) \log [p(\mathbf{y}|\mathbf{x})/p(\mathbf{y})]\, d\mathbf{y}$ in the continuous case the amount of information that a specific $\mathbf{x}$ gives about $\mathbf{y}$ (see Prob. 9-4), since its average over all $\mathbf{x}$ is $I(\mathbf{X};\mathbf{Y})$. Using (8-9), show that $J(x;Y) \geq 0$, and show that this quantity is invariant under one-to-one coordinate transformations but that it does not possess the additive property (8-20) which a measure of information ought to have. The fact that $J(x;Y) = H(Y) - H(Y|x)$ in the case of Prob. 9-4 is merely a result of the uniform prior distribution of $y$ and has no further significance. When is $J(x;Y) = 0$?

**Problem 9-9.** Suppose that the components of $\mathbf{x}$ are $m$ successive terms of an ergodic sequence of continuously distributed values. By dividing the $\mathbf{x}$ space into discrete cells of volume $dx^m$, show that the entropy of the resulting discrete distribution is $m \log dx$ less than the entropy of the original continuous distribution, and use (8-29) to show that, for any positive $\delta$ and $\epsilon$, the probability exceeds $1 - \delta$ that $p(\mathbf{x})$ lies between $b^{-H(\mathbf{X}) \pm \epsilon m}$ for sufficiently large $m$. Use (8-30) to show that the smallest volume in the $\mathbf{x}$ space containing a total probability $1 - \delta$ lies between $b^{H(\mathbf{X}) \pm \epsilon m}$ for sufficiently large $m$.

## □ THE BAND-LIMITED CHANNEL

In the case of a discrete communication channel the input and output are well-ordered sequences of distinct values, and it is natural to evaluate entropy and information rates in, say, bits per symbol, i.e., per value in the sequence. However, when the input and output are continuous functions of time, they may *always* be taking new values, and we cannot measure rates per "new value." Moreover, each "new value" will be highly correlated with the immediately preceding values and hence conveys a vanishing amount of entropy or information.

This difficulty can be overcome by means of the sampling theorem if, as we shall henceforth suppose, the waveforms are confined to frequencies less than $W$ cycles per second. As (7-6) shows, they are then determined completely by their values at intervals of $1/2W$, and we can express entropy and information rates in terms of *these* values just as in the discrete case.

We suppose that noise components of frequencies above $W$ are nearly but not completely absent, so that they affect the values of the received waveforms negligibly but limit the interval over which they can be accurately extrapolated. Thus, for example, $x(t)$ for $0 \leq t \leq T$ with $WT \gg 1$ (except near the ends of this interval) determines and is determined by approximately $2WT$ of its values $x_i = x(i/2W)$; i.e., it can be represented by a $2WT$-dimensional vector $\mathbf{x}$ with components $x_i$. As we saw in Chap. 7, the components of $\mathbf{x}$ in a suitably rotated coordinate frame are $\sqrt{2WT}$ times the amplitudes of the sinusoidal and cosinusoidal components into which $x(t)$ is resolved by Fourier series analysis covering the interval $(0,T)$.

**The Effect of Linear Filtering on Entropy Rate.** When such a waveform is passed through a linear filter, each of these frequency components is multiplied by the voltage attenuation factor $|G(f_j)|$ characterizing the response of the filter to its frequency. In addition, the filter's phase shift at any frequency effects a rotation in the plane representing the sinusoidal and cosinusoidal components of that frequency. Such rotations do not affect the volume of a small cube in the $\mathbf{x}$ space, but the attenuation $|G(f_j)|$ acts on each sinusoidal and cosinusoidal component, shrinking the volume of the cube by the factor

$$J = \prod_{j=1}^{WT} |G(f_j)|^2.$$

Since this jacobian is constant, the entropy change effected by the filter is simply its logarithm, which, as $T \to \infty$ and

$$f_{j+1} - f_j = df = \frac{1}{T} \to 0,$$

becomes

$$\log J = T \int_0^W \log |G(f)|^2 \, df.$$

Hence, dividing by $T$, we see that the filter increases the entropy rate per unit time of its input by

$$\int_0^W \log |G(f)|^2 \, df. \tag{9-8}$$

Although there may be some question regarding our treatment of the zero-frequency Fourier component, its contribution to (9-8) disappears after dividing by $T$ and passing to the limit.

## □ THE CAPACITY OF A CHANNEL WITH ADDITIVE NOISE

The input $x(t)$ and output $y(t)$ of a channel of bandwidth $W$ during a long time $T$ can be represented by the $2WT$-dimensional vectors $\mathbf{x}$ and $\mathbf{y}$ with components $x(i/2W)$ and $y(i/2W)$ for integer $i$, respectively. We now suppose that $y(t) = x(t) + n(t)$ or, equivalently,

$$\mathbf{y} = \mathbf{x} + \mathbf{n},$$

the noise $n(t)$, represented by the vector $\mathbf{n}$, being an ergodic random process statistically independent of the channel input $x(t)$ and, like $x(t)$ and $y(t)$, containing no frequency higher than $W$ (or hardly any).

Since the statistics of $\mathbf{y}$ for a fixed $\mathbf{x}$ are the same as those of $\mathbf{n}$ except for the displacement by $\mathbf{x}$, the prevarication $H(\mathbf{Y}|\mathbf{X}) = H(\mathbf{Y}|\mathbf{x})$ is identical with the entropy of the noise $H(\mathbf{N})$. Hence the mutual information of $\mathbf{x}$ and $\mathbf{y}$ is

$$\begin{aligned} I(\mathbf{X};\mathbf{Y}) &= H(\mathbf{Y}) - H(\mathbf{Y}|\mathbf{X}) \\ &= H(\mathbf{Y}) - H(\mathbf{N}). \end{aligned}$$

Dividing by $T$ and passing to the limit $T \to \infty$, we thus find that the channel's mutual information rate $I'(\mathbf{X};\mathbf{Y})$ is the difference between the entropy rate of its output, $H'(\mathbf{Y})$, and that of the noise, $H'(\mathbf{N})$.

Hence the channel capacity, the least upper bound of $I'(\mathbf{X};\mathbf{Y})$ for all acceptable input statistics $\mathbf{p}(\mathbf{x})$, is

$$C = \sup_{\mathbf{p}(\mathbf{x})} H'(\mathbf{Y}) - H'(\mathbf{N}). \tag{9-9}$$

Since $H'(\mathbf{N})$ is independent of $\mathbf{p}(\mathbf{x})$, it is attained by the input statistics that maximize the output entropy rate $H'(\mathbf{Y})$ subject to any restrictions that the channel imposes on $x(t)$ apart from the bandwidth limitation.

**Problem 9-10.** By applying (8-9) to $\int p(\mathbf{y}) \log \frac{q(\mathbf{y})}{p(\mathbf{y})}\, d\mathbf{y}$, where $p(\mathbf{y})$ and $q(\mathbf{y})$ are both probability density functions, show that

$$-\int p(\mathbf{y}) \log p(\mathbf{y})\, dy \leq -\int p(\mathbf{y}) \log q(\mathbf{y})\, d\mathbf{y},$$

with equality only when $q(\mathbf{y}) = p(\mathbf{y})$ for all $\mathbf{y}$.

**Problem 9-11.** Letting $q(\mathbf{y}) = (2\pi Q)^{-WT} \exp_e(-\tilde{\mathbf{y}}\mathbf{y}/2Q)$ in Prob. 9-10, where $\tilde{\mathbf{y}}\mathbf{y}$ is the square of the length of the $2WT$-component vector $\mathbf{y}$, show that, if the average power $\int \tilde{\mathbf{y}}\mathbf{y}p(y)\,dy/2WT$ of $y(t)$ is no more than $Q$, we have $H(\mathbf{Y}) \leq WT \log(2\pi eQ)$, with equality only when $y(t)$ has the statistics of white gaussian noise, described by the foregoing $q(\mathbf{y})$.

**White Gaussian Noise.** If the channel input is restricted to an average power no greater than $S$ and the noise has average power $N$, the average power of the channel output cannot exceed $S + N$ and, as Prob. 9-11 shows, the output entropy rate $H'(\mathbf{Y})$ cannot exceed $W \log 2\pi e(S + N)$. When the noise is white and gaussian, this upper bound can be attained by giving the input the statistics of white gaussian noise, for the output $y(t) = x(t) + n(t)$ will consequently have the same statistics, too.

Thus, substituting this upper bound into (9-9) and subtracting $H'(\mathbf{N}) = W \log(2\pi eN)$, which results from an entropy $\frac{1}{2} \log(2\pi eN)$ associated with each of the $2WT$ statistically independent components of $\mathbf{n}$ (Probs. 9-1 and 9-5), we get the channel capacity

$$C = W \log \frac{S + N}{N}. \tag{9-10}$$

This is the celebrated formula of Claude Shannon for the capacity of a channel of bandwidth $W$ that accepts signals of power no greater than $S$ as input, adding to them white gaussian noise of power $N$. Although our derivation dealt with the band from 0 to $W$ cps, including negative frequencies with the positive, the possibility of heterodyning any other band of equal width down to this band and of reversing this frequency translation shows that it applies to *any* band of width $W$. This can also be proved directly by dealing with the Fourier-coefficient components of the vectors or simply by noticing that (9-10) is proportional to $W$.

**Problem 9-12.** Taking into account the fact that the average noise power $N = N_0W$ is proportional to the channel bandwidth in the case of noise of constant power spectral density $N_0$, show that the channel capacity is zero for $W = 0$ and increases monotonically with $W$, approaching the limit $(S/N_0) \log e$ as $W \to \infty$.

**Realizing the Channel Capacity.** By dividing the $2WT$-dimensional signal space into discrete cells and letting the dimensions of these cells go to zero, the fundamental theorem on the noisy channel can be shown to apply to the continuous as well as to the discrete channel. Thus, for sufficiently large $T$, the reliable transmission of messages is likely to be achieved for any information rate $R < C$ by using statistically independent outputs of duration $T$ from a source having the statistics $\mathbf{p}(\mathbf{x})$ that maximize $H'(\mathbf{Y}) - H'(\mathbf{Y}|\mathbf{X})$ to represent $b^{RT}$ different messages of this duration.

In the case of (9-10), this is a white gaussian noise source of bandwidth $W$ and average power $S$, and each message will be represented by a channel input signal which is a random sample of such white gaussian noise. A catalog of these signals as used by the transmitter is made available to the receiver, which interprets the channel output $\mathbf{y}$ as that signal $\mathbf{x}$ which has the greatest posterior probability. If the $b^{RT}$ messages have equal prior probabilities, the message with the greatest posterior probability is the one represented by the signal $\mathbf{x}$ for which the likelihood $p(\mathbf{y}|\mathbf{x}) = (2\pi N)^{-WT} \exp_e [-(\tilde{\mathbf{y}} - \tilde{\mathbf{x}})(\mathbf{y} - \mathbf{x})/2N]$ is greatest, i.e., the $\mathbf{x}$ in the catalog for which $(\tilde{\mathbf{y}} - \tilde{\mathbf{x}})(\mathbf{y} - \mathbf{x}) = 2W\int [y(t) - x(t)]^2\, dt$ is least and which is thus the closest[1] to the received signal $\mathbf{y}$.

Because of the ergodicity of white gaussian noise, the time-averaged power of a random sample of duration $T$ tends to be very close to the ensemble average $S$ for large $T$. If each signal is multiplied by a constant to make its total energy *exactly* $ST$, then $\tilde{\mathbf{x}}\mathbf{x} = 2W\int x^2(t)\, dt$ will be the same for every signal, and the receiver's decisions can be based on maximizing $\tilde{\mathbf{x}}\mathbf{y} = 2W\int x(t)y(t)\, dt$, i.e., on cross-correlation.

This equalizing of the signal energies confines the signals $\mathbf{x}$ to the surface of a sphere of radius $\sqrt{2STW}$ centered at the origin. That it does not reduce $I'(\mathbf{X};\mathbf{Y}) = H'(\mathbf{Y}) - H'(\mathbf{N})$ below $C$ can be seen by noticing that, in either case, $y(t)$ has the same average power $S + N$, and $\mathbf{y}$ therefore tends to lie very close to the surface of a concentric sphere of radius $\sqrt{2(S + N)TW}$ and is uniformly distributed in direction. Thus the least volume containing a total probability $1 - \delta$ (Prob. 9-9) is changed in only one out of $2WT$ dimensions, viz., the thickness of the spherical shell, and the entropy *rate* $H'(\mathbf{Y})$ remains the same. [This thickness will be of the order of the standard deviation of the length of $\mathbf{y}$, which becomes $\sqrt{\frac{1}{2}(S + N)}$ when $\mathbf{x}$ is gaussian (see Prob. 9-15 on p. 195) and $\sqrt{\frac{1}{2}(2S + N)N/(S + N)}$ when $\mathbf{x}$ has length $\sqrt{2STW}$; in either case it is of the order of unity as $WT \to \infty$.]

**Colored Gaussian Noise.** If the noise added by the channel is gaussian but not white, i.e., if its power spectral density is a function $N_0(f)$ of frequency, and if the power spectral density of the channel input can be no more than $S_0(f)$, with[2]

$$S = \int_0^\infty S_0(f)\, df, \tag{9-11}$$

we can divide the spectrum into consecutive bands of width $df$, to each of which (9-10) applies. Since the noises in different bands are statistically

[1] V. A. Kotel'nikov, *loc. cit.*

[2] Note that power spectral densities denoted by $\Psi(f)$ in this book cover both negative and positive frequencies, but those denoted by other letters cover only the latter.

independent (see p. 141), their combined channel capacity is

$$C = \int_0^\infty \log \frac{S_0(f) + N_0(f)}{N_0(f)} \, df. \tag{9-12}$$

**The Best Signal Spectrum.** Differentiating this integrand with respect to $S_0(f)$, we get

$$\frac{\log e}{S_0(f) + N_0(f)}. \tag{9-13}$$

Hence, if $S_0(f)$ is arbitrary and is subject only to the restriction that it be nonnegative and that its integral, the total signal power (9-11), be $S$, and if (9-13) is not the same at all frequencies for which $S_0(f) > 0$ and as small or smaller at all other frequencies, $C$ can be increased by taking some of the signal power away from frequencies at which (9-13) is smaller and putting it on frequencies at which (9-13) is larger. Thus, subject to (9-11), (9-12) is greatest when the signal spectrum is such that $S_0(f) + N_0(f)$ has the same value $K$, chosen to satisfy (9-11), at all frequencies for which $S_0(f) > 0$ and an equal or larger value at all other frequencies. At the latter frequencies the spectral density of the noise is so great that they are not worth trying to use—unless $S$ is increased. Thus, for a given noise power spectral density $N_0(f)$, the best signal power spectral density is $K - N_0(f)$ where $N_0(f) \leq K$ and zero elsewhere.

**The Worst Noise Spectrum.** The derivative of the integrand in (9-12) with respect to $N_0(f)$ is

$$\frac{\log e}{S_0(f) + N_0(f)} - \frac{\log e}{N_0(f)}. \tag{9-14}$$

The nonnegative $N_0(f)$ giving total noise power

$$\int_0^\infty N_0(f) \, df = N \tag{9-15}$$

that minimizes (9-12) will be such as to make (9-14) the same at all frequencies for which $S_0(f) > 0$, since otherwise $C$ could be reduced by shifting noise power from a frequency at which (9-14) is larger to one at which it is smaller, i.e., more negative. Where $S_0(f) = 0$, for example, for $f > W$, (9-14) is zero, and the worst $N_0(f)$ is likewise zero.

When the signal spectrum is as good as possible, and the noise spectrum is simultaneously as bad as possible, (9-13) and (9-14) are both constant over the band to which the signal spectrum is confined. In this case the noise and signal spectra are both flat over this band, and the channel capacity $C$ attains a saddle-point;[1] i.e., any departure of

[1] N. M. Blachman, Communication as a Game, *IRE WESCON Convention Record:* Part 2, 61–66 (1957).

$S_0(f)$ from constancy would permit the channel capacity to be reduced below (9-10), and any departure of $N_0(f)$ from constancy would increase the channel capacity beyond (9-10).

Hence, for given signal and noise powers (9-11) and (9-15), the worst kind of additive gaussian noise is white gaussian noise. We shall shortly see that this is also the worst among *all* kinds of additive interference.

## ☐ NONGAUSSIAN NOISE

The problem of determining the channel capacity when the noise is not gaussian is, in general, unsolved. However, by introducing the concept of *entropy power* we shall be able to obtain useful upper and lower bounds for the channel capacity.

**Entropy Power.** As we have seen, white gaussian noise of bandwidth $W$ and power $\mathfrak{N}$ has entropy rate $H' = W \log 2\pi e\mathfrak{N}$. We shall define the "entropy power" of a band-limited random process as the power of the white gaussian noise having the same entropy rate. Hence, if its entropy rate is $H'$, its entropy power is

$$\mathfrak{N} = \frac{\exp_b (H'/W)}{2\pi e}. \tag{9-16}$$

Because (Prob. 9-11) white gaussian noise has the greatest possible entropy rate for a given average power $N$, we have

$$\mathfrak{N} \leq N, \tag{9-17}$$

with equality only for white gaussian noise.

If $n(t)$ is uncorrelated with $x(t)$ at the same instant, the variance of $y(t) = x(t) + n(t)$ is

$$\text{var }\{y(t)\} = \text{var }\{x(t)\} + \text{var }\{n(t)\}.$$

Since the average power of $x(t)$ is the sum of its variance plus the square of its mean, we have var $\{x(t)\} \leq S$. Similarly, var $\{n(t)\}$ cannot exceed the noise power $N$, and hence var $\{y(t)\} \leq S + N$. Because any d-c component (nonzero mean) of $y(t)$ cannot affect its entropy rate, by (9-17) its entropy power, which we shall call $\mathcal{P}$, cannot exceed its variance, and we have

$$\mathcal{P} \leq S + N, \tag{9-18}$$

with equality only when $x(t)$ and $n(t)$ both have mean zero and $y(t)$ is white gaussian noise. With $x(t)$ and $n(t)$ statistically independent, this condition can be fulfilled only if $x(t)$ and $n(t)$ too are gaussian.

If $x(t)$ and $n(t)$ are statistically independent of each other, and if $\mathcal{S}$ and $\mathfrak{N}$ are their entropy powers, it can be shown that

$$\mathcal{P} \geq \mathcal{S} + \mathfrak{N}, \tag{9-19}$$

with equality only when $x(t)$, $n(t)$, and, therefore, $y(t)$ are gaussian random processes with proportional power spectral densities. The proof[1] involves a long chain of inequalities for which we would have no other use, and we therefore omit it.

**Problem 9-13.** Show that, if a band-limited random process with entropy power $\mathfrak{N}$ is passed through a linear filter having frequency response $G(f)$, the entropy power of its output will be $\mathfrak{N} \exp \left( \int_0^W \log |G(f)|^2 \, df/W \right)$, this integral being the average, over the frequency band, of the attenuation expressed in logarithmic units (decibels for $b = 10^{0.1}$).

**Channel-Capacity Bounds.** In terms of entropy powers, the capacity (9-9) of a band-limited channel perturbed by additive noise can be written

$$C = \sup_{\mathbf{p}(\mathbf{x})} W \log \frac{\mathcal{P}}{\mathfrak{N}}.$$

Using (9-18) and (9-19) now, we get upper and lower bounds for $C$. For a given input power $S$, the lower bound is greatest when $\mathcal{S}$ is greatest, i.e., when $x(t)$ has the statistics of white gaussian noise and $\mathcal{S} = S$. Thus,

$$W \log \frac{S + \mathfrak{N}}{\mathfrak{N}} \leq C \leq W \log \frac{S + N}{\mathfrak{N}}, \tag{9-20}$$

the upper bound being attained only when the noise is gaussian and its spectrum is such that it can be flattened (made white) by the addition of a signal of power $S$. The lower bound is attained only for white gaussian noise, and in this case the upper and lower bounds are equal. Notice that, when $x(t)$ has white gaussian statistics, $I'(\mathbf{X};\mathbf{Y}) = W \log \mathcal{P}/\mathfrak{N}$ lies between these bounds.

**Problem 9-14.** (*a*) Show that the entropy power of gaussian noise with power spectral density $N_0(f)$ and bandwidth $W$ is $\mathfrak{N} = W \exp \left( \int_0^W \log N_0(f) \, df/W \right)$, which is $W$ times the geometric mean of $N_0(f)$, and verify (9-20) for this noise. (*b*) For given values of $W$, $S$, and $N$, sketch the part of

[1] A. J. Stam, "Some Mathematical Properties of Quantities of information," sec. 4.4, p. 102, Uitgeverij Excelsior, the Hague, April, 1959; A. J. Stam, Some Inequalities Satisfied by the Quantities of Information of Fisher and Shannon, *Information and Control,* **2**:101–112 (1959); N. M. Blachman, The Convolution Inequality for Entropy Powers, *IEEE Trans. Inform. Theory,* **IT-11**:267–271 (1965).

the $\mathfrak{N}C$ plane satisfying (9-17) and (9-20). Notice that, if $\mathfrak{N} = N$, then $C$ is uniquely restricted to its least possible value, (9-10).

**The Worst Noise.** For given values of $S$ and $N$, both bounds of (9-20) are least when $\mathfrak{N}$ is greatest, i.e., when $\mathfrak{N} = N$. (See Prob. 9-14*b*.) Since the two bounds are then equal, it follows that white gaussian noise minimizes the channel capacity and is thus the worst among *all* kinds of additive noise, gaussian and nongaussian.

**The Importance of the Channel Capacity.** Reliable communication through a noisy channel at rates approaching the channel capacity in general requires complicated coding and decoding. Both the information rates and the reliabilities of actual communication systems fall short of the ideal, but comparison of their information transmission rates with the channel capacity will indicate how much channel capacity is being wasted, and the $\mathbf{P}(\mathbf{x})$ for which $I'(\mathbf{X};\mathbf{Y}) = C$ may suggest ways to increase the information rate as well as the reliability. When the information rate is close to $C$, however, efforts to reduce the difference may well require an unwarranted increase in the complexity of the transmitter and receiver—not only to handle the optimum waveforms but also to store messages and received waveforms for a sufficient coding block length $T$ to yield reliable transmission.

In the next section we shall evaluate the error probability for the band-limited channel with additive white gaussian noise when the information rate is close to the channel capacity. From the result, the necessary block length can be estimated.

## □ GEOMETRIC TREATMENT OF THE CASE OF ADDITIVE WHITE GAUSSIAN NOISE

Our derivation, in terms of entropy rates, of the formula (9-10) for the capacity of a band-limited channel perturbed by additive white gaussian noise with specified input power is based on Shannon's 1948 paper.[1] In 1949 he published[2] an entirely independent derivation which made no use of the concept of entropy but, instead, used geometric methods to establish the limits of reliable transmission. Both approaches had been developed during World War II, the former stimulated by work on cryptography[3] (see p. 170). Because the geometric approach clarifies

[1] C. E. Shannon, A Mathematical Theory of Communication, *Bell System Tech. J.*, **27**:623–656 (1948).

[2] C. E. Shannon, Communication in the Presence of Noise, *Proc. IRE*, **37**:10–21 (1949).

[3] C. E. Shannon, Communication Theory of Secrecy Systems, *Bell System Tech. J.*, **28**:656–715 (1949).

the fundamental theorem on the noisy channel and facilitates the calculation of the error probability, we now take it up.

**The Noise Sphere.** Since all of the components of the vector **n** representing the additive white gaussian noise have mean zero and variance $N$ and are uncorrelated and jointly normal, the distribution of **n** is spherically symmetric about the origin. Hence, for any given transmitted signal **x**, the received signal **y** is distributed with spherical symmetry about **x** in a manner that does not depend on **x**. Because the $2WT$ components of **n** are statistically independent and identically distributed

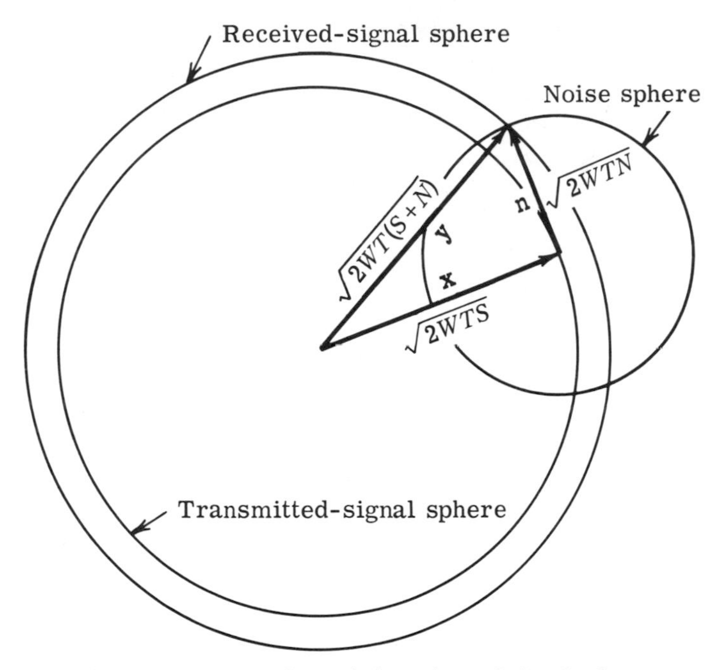


*Figure 9-1.* The vectors **x**, **n**, and **y** and the noise and signal spheres.

random variables, the sum of their squares, which is the square of the length of **n**, tends for large $T$ to be (by the law of large numbers) relatively close to $2WT$ times the mean-squared value $N$ of one component. Hence the length of **n** and, therefore, the distance between **y** and **x** tend to be very nearly $\sqrt{2WTN}$ (see Probs. 9-15 and 9-16). On the other hand, $\mathbf{n} = \mathbf{y} - \mathbf{x}$ is distributed uniformly in direction (Prob. 7-8).

Thus we may imagine the transmitted-signal vector **x** to be surrounded by a "noise sphere" of radius $\sqrt{2WTN}$, the received-signal vector **y** being uniformly distributed over its surface or, more precisely, *near* its surface (Fig. 9-1). If $S$ is the average power of the transmitted signal, the length of **x** must be very close to $\sqrt{2WTS}$ because of (7-8).

Since $S + N$ is the average power of the received signal $y(t) = x(t) + n(t)$, the length of $\mathbf{y}$ is likewise very close to $\sqrt{2WT(S + N)}$.

From the fact that this is the square root of the sum of the squares of the (approximate) lengths of $\mathbf{x}$ and of $\mathbf{n}$, it follows by the pythagorean theorem that $\mathbf{x}$ and $\mathbf{n}$ are very nearly perpendicular (see Prob. 9-17). This conclusion does not contradict the assertion above that $\mathbf{y}$ is uniformly distributed over the noise sphere, for nearly all of the surface of this sphere lies very nearly 90° away from any given point on the surface because of its high dimensionality. Although this is clearly not the case in two dimensions, it begins to be true in three, where the area near the equator considerably exceeds that near the poles. The foregoing conclusion is supported by the observation that the component of $\mathbf{n}$ parallel to $\mathbf{x}$ on the average represents only $1/2WT$ of the total noise energy, the remainder being represented by the other $2WT - 1$ components, all orthogonal to $\mathbf{x}$.

**Problem 9-15.** From the fact (Prob. 7-8) that the standard deviation of the square of the length of the noise vector $\mathbf{n}$ is $1/\sqrt{WT}$ times its mean, show that the standard deviation of the length itself is $\sqrt{\frac{1}{2}N}$ for $WT \gg 1$. Hence, with high probability, the length of $\mathbf{n}$ lies within a few times $\sqrt{\frac{1}{2}N}$ of $\sqrt{2WTN}$.

**Problem 9-16.** Show that the mode of the distribution of the length of $\mathbf{n}$ (Prob. 7-8) is $\sqrt{(2WT - 1)N}$. By comparing the second derivative of the logarithm of the probability density function at the mode with the second derivative of a normal probability density function having the same mode, show that it is approximated by a normal distribution with standard deviation $\sqrt{\frac{1}{2}N}$.

**Problem 9-17.** Show that the standard deviation of the angle between $\mathbf{x}$ and $\mathbf{n}$ is $1/\sqrt{2WT}$ for $WT \gg 1$.

**The Maximum Rate of Reliable Transmission.** In the absence of any restriction on the signals $\mathbf{x}$, it is possible to choose an arbitrarily large number of them with nonoverlapping noise spheres and, hence, to transmit information reliably at any rate, however large. The infinite channel capacity in this case results from the fact that the effect of the noise is small compared to the unbounded available separation of the input signals.

A convenient and not unrealistic restriction on the input signals is our assumption that their average power does not exceed $S$. With $x(t)$ ergodic, $\mathbf{x}$ is thus closely restricted to a sphere of radius $\sqrt{2WTS}$ about the origin for $WT \gg 1$. Consequently, the average power of the output signal $y(t)$ is no more than $S + N$, and $\mathbf{y}$ is closely confined to a sphere of radius $\sqrt{2WT(S + N)}$ about the origin (Fig. 9-1).

Reliable communication does not require the noise spheres surrounding the signal vectors representing different messages not to overlap. (See the geometric interpretation in the case of binary signaling in Chap. 7, where each of the two noise spheres may contain the center of the other without difficulty.) Rather, it requires that the part of each noise sphere which does not overlap any other noise sphere represent nearly the entire volume of that noise sphere, so that $\mathbf{y}$ is overwhelmingly likely to lie in the noise sphere of the $\mathbf{x}$ that was sent and in no other.

Since at least half the volume of a noise sphere with its center in the signal sphere of radius $\sqrt{2WTS}$ lies within the received-signal sphere of radius $\sqrt{2WT(S + N)}$, the number of reliably distinguishable signals cannot exceed twice the ratio of the volume of the latter to that of a noise sphere. This ratio is the $2WT$th power of the ratio of the radii, $\sqrt{(S + N)/N}$. Hence there can be no more than $2(1 + S/N)^{WT}$ reliably distinguishable signals. In writing this expression we have neglected a factor whose logarithm is $O(\sqrt{WT})$, i.e., of the order of $\sqrt{WT}$, which results from the fact that the noise vector can extend a few standard deviations beyond the surface of the noise sphere. However, this factor disappears from the limit we are about to take.

Thus the logarithm of the number of reliably distinguishable signals cannot exceed $WT \log (1 + S/N) + \log 2 + O(\sqrt{WT})$. By (8-10) this is an upper bound on the amount of information that can be reliably conveyed in a time $T$. Dividing by $T$ and passing to the limit $T \to \infty$, we see that the rate of reliable transmission of information cannot exceed $W \log (1 + S/N)$. As in the proof of the fundamental theorem on the noisy channel, we now use random coding to show that this upper bound can actually be approached arbitrarily closely.

If during a time $T$ the channel is to convey one of $m$ different equiprobable messages, these may be represented by any $m$ different signals with the corresponding vectors in or on the signal sphere, though some choices will result in higher error probabilities than others. We shall suppose that they are chosen randomly and independently from a uniform distribution over the surface of the signal sphere (see p. 189), and we shall show that the resulting error probability, averaged over all such choices, will be small for large $T$ if $m$ is not too large.

If the transmitted signal $\mathbf{x}$ is the only signal within a distance $\sqrt{2WTN}$ of the received signal $\mathbf{y}$, no error will be made. The probability that this is the case is the probability that none of the $m - 1$ other signals lie in the spherical cap cut out of the signal sphere by a sphere of radius $\sqrt{2WTN}$ about a point $\mathbf{y}$ on the received-signal sphere (Fig. 9-2). The probability that any one of the $m - 1$ lies in this cap is the ratio of the area of the cap to the area of the signal sphere. The area of the cap does not exceed the area of a sphere of equal diameter.

Making use of the fact that, in similar (right) triangles, the ratios of corresponding sides are equal, we find that the radius of this sphere of equal diameter is $\sqrt{2WTSN/(S+N)}$. The ratio of its area to that of the signal sphere is therefore $(1 + S/N)^{\frac{1}{2}-WT}$. Thus the probability that any one of the $m - 1$ other signals lies outside the cap is at least $1 -$

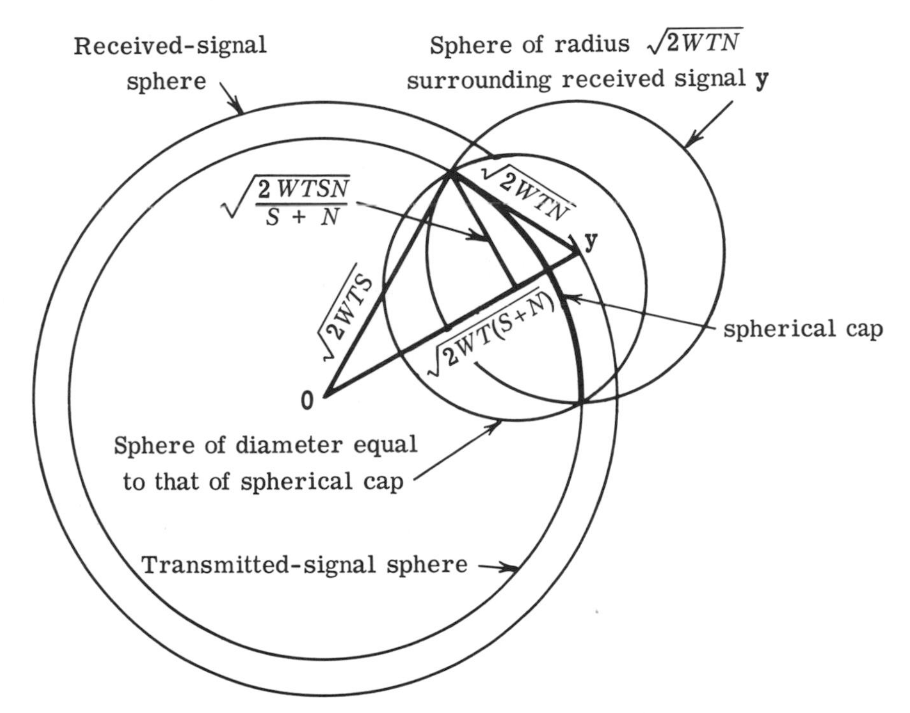


*Figure 9-2.* A two-dimensional cross section of the spheres and spherical cap, which are rotationally symmetric about the line **Oy**.

$(1 + S/N)^{\frac{1}{2}-WT}$, and the probability that all of them lie outside the cap is at least

$$\begin{aligned}[1 - (1 + S/N)^{\frac{1}{2}-WT}]^{m-1} &\geq 1 - (m-1)(1 + S/N)^{\frac{1}{2}-WT} \\ &\geq 1 - b^{RT}(1 + S/N)^{\frac{1}{2}-WT},\end{aligned}$$

where $R = (\log m)/T$ is the information rate of the message source, since, with $u < 1$, $(1 - u)^r$ is convex downward for fixed $r > 1$ and has tangent $1 - ru$ at $u = 0$. The error probability is therefore no more than

$$b^{RT}(1 + S/N)^{\frac{1}{2}-WT} = \sqrt{1 + S/N} \exp \{[R - W \log (1 + S/N)]T\},$$

which goes to zero as $T \to \infty$ if $R < W \log (1 + S/N)$. The least upper bound of the information rates satisfying this inequality is (9-10), $C = W \log (1 + S/N)$.

As we saw on p. 196, there is no sequence of codes for increasing $T$ that yields a vanishing error probability for any information rate exceeding this value. The fact that we made no use of the interior of the signal sphere for our random code did not prevent us from approaching this bound, because nearly all of the volume of a sphere lies within a distance of the surface equal to a few times the ratio of the radius to the dimensionality, and it is pre-empted by the signals on the surface. The use of the interior could therefore add relatively few reliably distinguishable signals.

As in the general proof of the fundamental theorem on the noisy channel, from the fact that the error probability averaged over all codes vanishes as $T \to \infty$, it follows that there exist codes yielding arbitrarily small error probabilities. Thus we have proved the fundamental theorem for the particular case of the band-limited channel with additive white gaussian noise and specified input power and have, at the same time, determined the supremum of the rates at which information can be reliably sent through it, which is the channel capacity (9-10) that we found earlier.

In the case of band-limited noise of power $N$ that is not white and gaussian, $\mathbf{n}$ is still confined closely to the sphere of radius $\sqrt{2WTN}$ about the origin representing waveforms of power $N$, but the smallest volume containing a total probability $1 - \delta$ is only a small fraction of the volume of the sphere (see Prob. 9-9), being equal to the volume of a sphere of radius $\sqrt{2WT\mathfrak{N}}$, where $\mathfrak{N}$ is its entropy power. Such noise yields a channel capacity larger than (9-10) because this smaller noise volume permits significantly more signals to be fitted into the signal sphere without serious overlapping of the volumes in which the corresponding received signals are likely to fall.

**Problem 9-18.** Show that, if $\mathbf{n}$ is uniformly distributed over the surface of a sphere, each of its $2WT$ components becomes normally distributed as $T \to \infty$, and any orthogonal pair of components is uncorrelated. Show further that the joint distribution of such a pair becomes normal and that any $k$ orthogonal components likewise become jointly normal.

**Dirichlet Regions.**† As we noted on p. 189, when all of the signals chosen to represent different messages have equal prior probability, and a signal $\mathbf{y}$ is received, it is most likely to represent the message corre-

† This section and the next (Error Probability) may be omitted on the first reading.

sponding to the nearest **x** among the chosen set of signals if the channel is perturbed by additive white gaussian noise. Associated with each such **x**, then, is the set of all possible received signals **y** that are closer to this **x** than to any other. This set is called the *Dirichlet region* of that **x**. In this way the entire space of received signals **y** is divided into mutually exclusive Dirichlet regions, the points on their boundaries which are equidistant from two or more signals being assigned arbitrarily to the adjoining regions.

These regions thus decode the received signals optimally, as they show the most likely **x** for each **y**, thereby minimizing the error probability. When $m = \exp(RT)$ signals are distributed uniformly over the

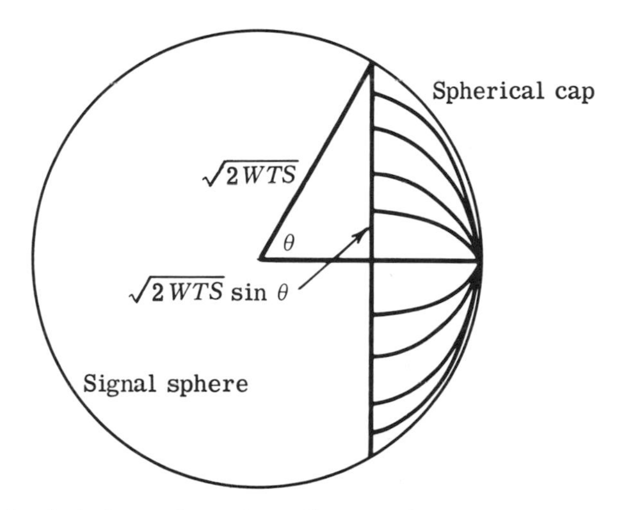


*Figure 9-3.* A spherical cap of angular radius $\theta$ on the signal sphere.

surface of the signal sphere, the number that can be expected to fall within a spherical cap (Fig. 9-3) of angular radius $\theta$ (subtended at the center of the sphere) is $\exp(RT)$ times the ratio of the area of such a cap to that of the sphere. Since the area of the cap is less than that of a sphere of the same diameter, this ratio is less than $\sin^{2WT-1}\theta$. Hence, when $\theta <$ arcsin exp $(-R/2W)$, as $T \to \infty$ the expected number of signals falling within the cap goes to zero. In fact, this number will go to zero whenever arcsin exp $(-R/2W)$ exceeds $\theta$ by a large multiple of $1/WT$.

On the other hand, the area of a unit sphere is its dimensionality times its volume, which does not exceed twice the area of its equatorial disk. Hence the area of the signal sphere does not exceed $4WT$ times the area of a disk of the same diameter. (The area of the signal sphere is actually approximately $2\sqrt{\pi WT}$ times the area of its equatorial disk; see Prob. 7-9.) Since the area of the cap is at least the area of a flat circular disk of the same diameter, whenever $\theta > \arcsin \exp(-R/2W)$ the expected number of signals falling within the cap becomes very large for $WT \gg 1$. More precisely, the expected number of signals falling within the cap will be large whenever $\theta$ exceeds $\arcsin \exp(-R/2W)$ by a large multiple of $(\log_e WT)/WT$. Thus the transition from a very small expectation to a very large one occurs within a very small neighborhood of

$$\theta = \arcsin \exp(-R/2W). \tag{9-21}$$

The results of Probs. 9-15 and 9-17 show that the width of this neighborhood, of the order of $(\log_e WT)/WT$, is smaller than the dispersion of the angle between **x** and **y** due to the noise, which is of the order of $1/\sqrt{WT}$.

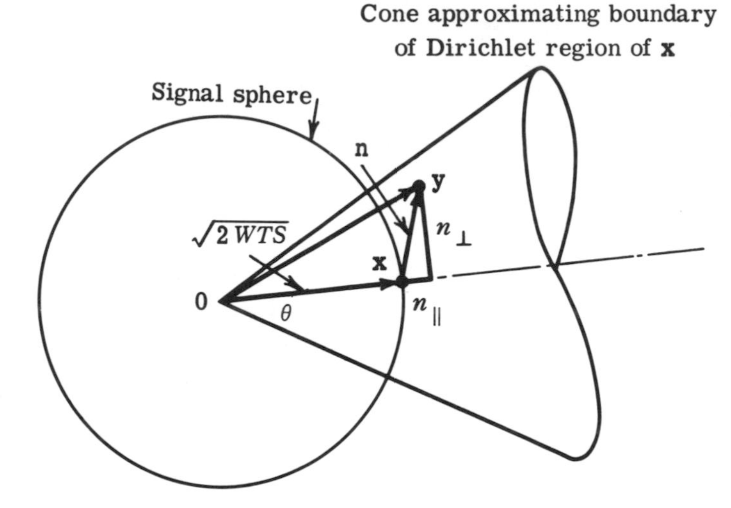


*Figure 9-4.* The signal **x**, the two components of the noise **n**, and the cone approximating the boundary of the Dirichlet region of **x**.

The boundaries of the Dirichlet regions are formed by the hyperplanes that are equidistant from two different signals. Because all of the signals are equidistant from the origin when they lie on the signal sphere, all of these hyperplanes pass through the origin, and the Dirichlet regions are all pyramids with apex at the origin. From (9-21) it follows that any signal's nearest neighbor is very nearly at angular distance $\theta = \arcsin \exp(-R/2W)$; hence the nearest face of its Dirichlet region is very nearly at angular distance[1] $\frac{1}{2} \arcsin \exp(-R/2W)$.

On the other hand, going in any direction on the signal sphere from a given signal, we do not reach the boundary of its Dirichlet region until

[1] N. M. Blachman, Some Properties of Large Sets of Random Signals, *Ann. Math. Statist.*, **32**:916 (1961).

the spherical cap centered on the point we are at, with radius equal to the distance to the given signal, contains another signal too. Hence in a random direction the boundary of the Dirichlet region is very nearly arcsin exp $(-R/2W)$ away from the given signal; i.e., in nearly every direction it is very close to the cone of half-angle (9-21) with axis through the given signal and apex at the origin (Fig. 9-4).

Although we shall make no use of the fact, it is interesting to note that nearly every signal within an angular distance of about twice (9-21) contributes a face to the Dirichlet region of a given signal. However, nearly all of the surface of the region is accounted for by faces that are approximately arcsin $[2^{-\frac{1}{2}} \exp(-R/2W)]$ from the signal, since a line (arc of a great circle) on the signal sphere in a random direction from this signal meets the boundary in a point such that the line joining it to the signal of the neighboring Dirichlet region is very nearly perpendicular to the original line (almost all of the area of a sphere of high dimensionality is nearly 90° away from any given point on the sphere), and the bounding hyperplane must bisect this angle. By the law of sines, a right spherical triangle with hypotenuse $\theta$ and a 45° angle has opposite side $\arcsin(2^{-\frac{1}{2}} \sin\theta) = \arcsin[2^{-\frac{1}{2}} \exp(-R/2W)]$.

**Problem 9-19.** Show that, although with randomly chosen signals the distance from nearly every signal to the nearest face of its Dirichlet region is close to $\frac{1}{2}\theta$, the smallest such distance among all of the signals is very close to $\frac{1}{2} \arcsin \sin^2\theta$. [HINT: Show that the expected number of pairs of signals separated by less than $\phi$ is $\frac{1}{2}\frac{S(\pi)}{S(\theta)}\left[\frac{S(\pi)}{S(\theta)} - 1\right]\frac{S(\phi)}{S(\pi)}$, where $S(\phi)$ is the area of a cap of angular radius $\phi$ on a unit sphere, and $S(\pi) = 2S(\frac{1}{2}\pi)$ is the area of the sphere, the dominant factor in $S(\phi)$ being $\sin^{2WT}\phi$ for $\phi \le \frac{1}{2}\pi$.]

**Error Probability.** The probability of error of the optimum receiver, which bases its decisions on the Dirichlet regions in the case of signals of equal prior probability to which the channel adds noise whose probability density function decreases monotonically with distance, is the probability that the received signal $\mathbf{y}$ will fall outside the Dirichlet region of the transmitted signal $\mathbf{x}$. For our randomly chosen signals, the average error probability, averaged over all possible choices of the signals, is therefore approximately the probability that $\mathbf{y}$ falls outside the cone of half-angle (9-21) with apex at the origin and axis passing through $\mathbf{x}$. Since the noise $\mathbf{n} = \mathbf{y} - \mathbf{x}$ has a spherically symmetric normal distribution, this probability is given by the noncentral $t$ distribution.[1]

[1] C. E. Shannon, Probability of Error for Optimal Codes in a Gaussian Channel, *Bell System Tech. J.*, **38**:611–656 (1959).

When the information rate $R$ is close to the channel capacity, this average error probability takes a relatively simple form, which we shall now find. We resolve the noise vector $\mathbf{n}$ into two components, one of (positive or negative) length $n_\parallel$ parallel to the signal vector $\mathbf{x}$ and the other of (positive) length $n_\perp$ perpendicular to $\mathbf{x}$ (Fig. 9-4); $n_\parallel$ is normal, with mean zero and variance $N$, and $n_\perp$ can be regarded as normally distributed with mean $\sqrt{2WTN}$ and variance $\frac{1}{2}N$ (Probs. 9-15 and 9-16) for deviations from the mean that are not too large, since the central limit theorem is applicable to $n_\perp{}^2$. From Fig. 9-4 we see that an error will be made if $n_\perp > (\sqrt{2WTS} + n_\parallel) \tan \theta$. For $R < C$ the probability of this event is, by (1-6),

$$\operatorname{erf} \frac{\sqrt{2WTN} - (\sqrt{2WTS} + n_\parallel) \tan \theta}{\sqrt{\frac{1}{2}N}} \cong \frac{\exp_e \left( - \dfrac{[\sqrt{2WTN} - (\sqrt{2WTS} - n_\parallel) \tan \theta]^2}{N} \right)}{\sqrt{8\pi WT}\,(\sqrt{S/N} \tan \theta - 1)}.$$

To average this error probability over the distribution of $n_\parallel$, we multiply by its probability density function, $[\exp_e(-\frac{1}{2}n_\parallel{}^2/N)]/\sqrt{2\pi N}$, and integrate from $-\infty$ to $\infty$ with respect to $n_\parallel$ by completing its square. Thus the average error probability is approximately

$$P = \frac{\exp_e \left( - \dfrac{2WT(\sqrt{S/N} \tan \theta - 1)^2}{1 + 2 \tan^2 \theta} \right)}{\sqrt{8\pi WT}\,\sqrt{1 + 2 \tan^2 \theta}\,(\sqrt{S/N} \tan \theta - 1)} \tag{9-22}$$

for $R < C$, provided that $T$ is large enough to justify our normal approximation to the distribution of $n_\perp$; the smaller $P$ is, the larger $T$ must be. As (1-8) indicates, the average error probability for $R > C$ is 1 plus (9-22), which is then negative.

When $R$ is close to $C$, $\tan^2 \theta$ is close to $N/S$, by (9-10) and (9-21), and we can use this approximation in (9-22) except in the parentheses, which would vanish. There we can instead write

$$\begin{aligned}
\sqrt{S/N} \tan \theta - 1 &= \frac{S/N - \cot^2 \theta}{\sqrt{S/N} \cot \theta + \cot^2 \theta} \\
&= \frac{\exp (C/W) - \exp (R/W)}{\sqrt{S/N} \cot \theta + \cot^2 \theta} \\
&= \frac{(1 + S/N)\{1 - \exp[-(C - R)/W]\}}{\sqrt{S/N} \cot \theta + \cot^2 \theta} \\
&\cong \frac{S + N}{S} \frac{C - R}{2W \log e}.
\end{aligned}$$

With these approximations, (9-22) becomes[1]

$$P = \sqrt{\frac{WS^3}{2\pi T(S + 2N)}} \frac{\exp\left[-\frac{(S+N)^2(C-R)^2T}{2WS(S+2N)\log e}\right]}{(S+N)(C-R)/\log e}. \tag{9-23}$$

The coefficient of $-T$ in the exponent here is called the "reliability."[2] The coding block length $T$ required to get a given small error probability $P$ is approximately the ratio of $-\log P$ to the reliability, since the exponential varies much more rapidly with $T$ than do the other factors in (9-23). As $R$ approaches $C$, the required block length becomes infinite.

Because (9-23) is the error probability for a random code, averaged over all such codes, there must be a code that is at least this good. To see that there can be no code with a significantly lower error probability, we notice that if, for the purpose of decoding, the Dirichlet region of a given signal is replaced by a cone enclosing the same solid angle with its axis through that signal, the probability of its correct decoding is increased, since the replacement involves substituting parts of the cone that are outside the pyramid for parts of the pyramid that are outside the cone, and the received signal has a higher probability density in the former than in the latter. If the cones around two different signals are unequal, the average error probability for these two signals can similarly be reduced by replacing each cone by another enclosing half the sum of the solid angles of the first two cones. Thus the error probability for the best code cannot be less than the value obtained by replacing the Dirichlet region of the transmitted signal by a cone enclosing $\exp(-RT) = 1/m$ of the signal sphere, i.e., by a cone of half-angle very close to (9-21). This, in fact, is how we obtained (9-23), although, because of our approximations, only its exponential factor is meaningful.

We cannot significantly reduce the error probability by making use of the interior of the signal sphere,[3] for, as we have noted earlier, nearly all of the volume of the sphere lies within a distance of the order of $\sqrt{S/WT}$ from its surface, and this distance is of a smaller order than the standard deviation $\sqrt{\frac{1}{2}N}$ of the length of the noise vector.

## □ THE INFORMATION RATE OF A CONTINUOUS SOURCE

In the continuous case, the information rate of a source appears to be infinite, since any alteration in its output, however slight, seems to repre-

[1] Shannon, *ibid.*, eq. (73). Equation (9) should be changed to agree with eq. (73).

[2] This definition is based on the fact that for a large variety of channels the error probability falls exponentially (as it does here) with increasing block length $T$ when $T$ is large and $R < C$. See C. E. Shannon, Certain Results in Coding Theory for Noisy Channels, *Information and Control*, **1**:6–25 (1957), and Gallager, *op. cit.*

[3] N. M. Blachman, The Closest Packing of Equal Spheres in a Larger Sphere, *Am. Math. Monthly*, **70**:526–529 (1963).

sent a different message. If, on the other hand, outputs that are nearly equal in some appropriate sense are regarded as representing the same message, we can define the information rate of the source relative to a "fidelity criterion."[1] For this purpose we introduce a measure $D(\mathbf{x};\mathbf{y})$ of the difference between the source output $x(t)$ and a distorted version of it, $y(t)$, which is to represent the same message, and we require the average of $D(\mathbf{x};\mathbf{y})$ over the joint distribution of $\mathbf{x}$ and $\mathbf{y}$ not to exceed some prescribed value $D$, which is the allowable average distortion.

This condition will, in general, be satisfied by many random processes $y(t)$ that are suitably dependent on $x(t)$, and we define the information rate $R$ of $x(t)$ subject to this fidelity criterion as the infimum (greatest lower bound) of the information rates $H'_d(\mathbf{Y})$ of these $y(t)$, where the subscripts $d$ and $c$ will refer to entropy rates as defined in the discrete and continuous cases, respectively. Since $H'_d(\mathbf{Y})$ is infinite whenever $y(t)$ has a continuous distribution, this minimum can be attained only when the distribution of $\mathbf{y}$ is discrete. Nevertheless, $R$ is given by

$$R = \inf_{\mathbf{p}(\mathbf{y}|\mathbf{x})} I'(\mathbf{X};\mathbf{Y}), \tag{9-24}$$

where the infimum is over all conditional distributions $\mathbf{p}(\mathbf{y}|\mathbf{x})$ for which $\mathrm{E}\,\{D(\mathbf{x};\mathbf{y})\} = \int\int p(\mathbf{x})p(\mathbf{y}|\mathbf{x})\, D(\mathbf{x};\mathbf{y})\, d\mathbf{y}\, d\mathbf{x} \leq D$, and the distribution of $\mathbf{y}$ may be continuous.

To prove (9-24), we first note that $R = \inf H'_d(\mathbf{Y})$ cannot be less than $\inf I'(\mathbf{X};\mathbf{Y}) = \inf\,[H'_c(\mathbf{Y}) - H'_c(\mathbf{Y}|\mathbf{X})] \leq \inf\,[H'_d(\mathbf{Y}) - H'_d(\mathbf{Y}|\mathbf{X})]$, since the continuous case includes as a limit the discrete case, where $H'_d(\mathbf{Y}|\mathbf{X}) \geq 0$. To show that $R$ is *equal* to $\inf I'(\mathbf{X};\mathbf{Y})$, we consider the conditional probability density function $p(\mathbf{y}|\mathbf{x})$ yielding this infimum. Determining $p(\mathbf{y}) = \int p(\mathbf{x})p(\mathbf{y}|\mathbf{x})\, d\mathbf{x}$ and dividing the $\mathbf{y}$ space into cells of volume $d\mathbf{y}$, we obtain a discrete distribution, from which we choose $b^{(I'+\epsilon)T}$ random values of $\mathbf{y}$ independently, where $\epsilon > 0$ and $I' = \inf I'(\mathbf{X};\mathbf{Y})$.

For a random $\mathbf{x}$, the distribution $p(\mathbf{y}|\mathbf{x})\, d\mathbf{y}$ gives a typical set of values of $\mathbf{y}$ satisfying an inequality like (8-29), and, as in the proof of the fundamental theorem on the noisy channel, as $T \to \infty$, this set will, with probability approaching 1, contain at least one of the chosen values of $\mathbf{y}$. Any one of these will serve as a deterministic, discrete representation of $\mathbf{x}$, and it will satisfy the fidelity criterion because the time statistics of this typical $\mathbf{y}$ will approximate the ensemble statistics $p(\mathbf{y}|\mathbf{x})\, d\mathbf{y}$. Since $H'_d(\mathbf{Y})$, by (8-10), is no more than $I' + \epsilon$, we have $R \leq I'$, and (9-24) is proved.

[1] C. E. Shannon, A Mathematical Theory of Communication, *Bell System Tech. J.*, **27**:623–656 (1948), Part V; C. E. Shannon, Coding Theorems for a Discrete Source with a Fidelity Criterion, in "Information and Decision Processes," R. E. Machol (ed.), pp. 93–126, McGraw-Hill Book Company, New York, 1960.

From (9-24) we see that a channel of capacity $C \geq R$ is necessary and a channel having any capacity $C > R$ is sufficient to permit the reproduction [namely, $y(t)$] of $x(t)$ to within the tolerance allowed by the fidelity criterion. If the message $y(t)$ is given to the transmitter, by the fundamental theorem on the noisy channel the receiver can reproduce $y(t)$ with a probability arbitrarily close to 1, thereby reproducing $x(t)$ satisfactorily. Thus the information rate of any source relative to a suitable fidelity criterion is the channel capacity needed for satisfactorily communicating the output of that information source.

**Problem 9-20.** Using the mean-squared difference $D(\mathbf{x};\mathbf{y}) = \int_0^T [x(t) - y(t)]^2 \, dt/T = (\tilde{\mathbf{x}} - \tilde{\mathbf{y}})(\mathbf{x} - \mathbf{y})/2WT$ as the measure of fidelity and requiring its expectation not to exceed $N$, show that $R \geq W \log \mathcal{Q}/N$, where $\mathcal{Q}$ is the entropy power of $x(t)$ and $W$ is its bandwidth, with equality only when $x(t)$ is the sum of white gaussian noise of power $N$ plus a statistically independent random process.

**Problem 9-21.** By choosing the values of $\mathbf{y}$ in the proof of (9-24) randomly over the surface of a sphere of radius $\sqrt{2WT(Q - N)}$ instead of optimally, show that, under the conditions of Prob. 9-20, $R \leq W \log Q/N$, where $Q$ is the mean-squared value of $x(t)$.

# INDEX

Italic page numbers refer to problems. Authors' names are followed by the numbers of only those pages with complete citations.

58863

DATE DUE